AF566762

High Pressure Technologies in Biomass Conversion

Green Chemistry Series

Editor-in-chief:
James H. Clark, *Department of Chemistry, University of York, UK*

Series editors:
George A. Kraus, *Iowa State University, USA*
Andrzej Stankiewicz, *Delft University of Technology, The Netherlands*
Peter Siedl, *Federal University of Rio de Janeiro, Brazil*

Titles in the series:
1: The Future of Glycerol: New Uses of a Versatile Raw Material
2: Alternative Solvents for Green Chemistry
3: Eco-Friendly Synthesis of Fine Chemicals
4: Sustainable Solutions for Modern Economies
5: Chemical Reactions and Processes under Flow Conditions
6: Radical Reactions in Aqueous Media
7: Aqueous Microwave Chemistry
8: The Future of Glycerol: 2nd Edition
9: Transportation Biofuels: Novel Pathways for the Production of Ethanol, Biogas and Biodiesel
10: Alternatives to Conventional Food Processing
11: Green Trends in Insect Control
12: A Handbook of Applied Biopolymer Technology: Synthesis, Degradation and Applications
13: Challenges in Green Analytical Chemistry
14: Advanced Oil Crop Biorefineries
15: Enantioselective Homogeneous Supported Catalysis
16: Natural Polymers Volume 1: Composites
17: Natural Polymers Volume 2: Nanocomposites
18: Integrated Forest Biorefineries
19: Sustainable Preparation of Metal Nanoparticles: Methods and Applications
20: Alternative Solvents for Green Chemistry: 2nd Edition
21: Natural Product Extraction: Principles and Applications
22: Element Recovery and Sustainability
23: Green Materials for Sustainable Water Remediation and Treatment
24: The Economic Utilisation of Food Co-Products
25: Biomass for Sustainable Applications: Pollution Remediation and Energy
26: From C-H to C-C Bonds: Cross-Dehydrogenative-Coupling
27: Renewable Resources for Biorefineries
28: Transition Metal Catalysis in Aerobic Alcohol Oxidation
29: Green Materials from Plant Oils
30: Polyhydroxyalkanoates (PHAs) Based Blends, Composites and Nanocomposites

31: Ball Milling Towards Green Synthesis: Applications, Projects, Challenges
32: Porous Carbon Materials from Sustainable Precursors
33: Heterogeneous Catalysis for Today's Challenges: Synthesis, Characterization and Applications
34: Chemical Biotechnology and Bioengineering
35: Microwave-Assisted Polymerization
36: Ionic Liquids in the Biorefinery Concept: Challenges and Perspectives
37: Starch-based Blends, Composites and Nanocomposites
38: Sustainable Catalysis: With Non-endangered Metals, Part 1
39: Sustainable Catalysis: With Non-endangered Metals, Part 2
40: Sustainable Catalysis: Without Metals or Other Endangered Elements, Part 1
41: Sustainable Catalysis: Without Metals or Other Endangered Elements, Part 2
42: Green Photo-active Nanomaterials
43: Commercializing Biobased Products: Opportunities, Challenges, Benefits, and Risks
44: Biomass Sugars for Non-Fuel Applications
45: White Biotechnology for Sustainable Chemistry
46: Green and Sustainable Medicinal Chemistry: Methods, Tools and Strategies for the 21st Century Pharmaceutical Industry
47: Alternative Energy Sources for Green Chemistry
48: High Pressure Technologies in Biomass Conversion

How to obtain future titles on publication:
A standing order plan is available for this series. A standing order will bring delivery of each new volume immediately on publication.

For further information please contact:
Book Sales Department, Royal Society of Chemistry, Thomas Graham House, Science Park, Milton Road, Cambridge, CB4 0WF, UK
Telephone: +44 (0)1223 420066, Fax: +44 (0)1223 420247
Email: booksales@rsc.org
Visit our website at www.rsc.org/books

High Pressure Technologies in Biomass Conversion

Edited by

Rafał M. Łukasik
Unidade de Bioenergia, Portugal
Email: rafal.lukasik@lneg.pt

Green Chemistry Series No. 48

Print ISBN: 978-1-78262-485-1
PDF eISBN: 978-1-78262-676-3
EPUB eISBN: 978-1-78801-129-7
ISSN: 1757-7039

A catalogue record for this book is available from the British Library

The Royal Society of Chemistry is a charity, registered in England and Wales, Number 207890, and a company incorporated in England by Royal Charter (Registered No. RC000524), registered office: Burlington House, Piccadilly, London W1J 0BA, UK, Telephone: +44 (0) 207 4378 6556.

For further information see our web site at www.rsc.org

Printed in the United Kingdom by CPI Group (UK) Ltd, Croydon, CR0 4YY, UK

FOREWORD

High Pressure Processing for the Valorization of Biomass

HÉCTOR A. RUIZ

Biorefinery Group, Food Research Department, School of Chemistry, Autonomous University of Coahuila, 25280, Saltillo, Coahuila, Mexico
Cluster of Bioalcoholes, Mexican Centre for Innovation in Bioenergy (Cemie-Bio), Mexico
Email: hector_ruiz_leza@uadec.edu.mx

Due to environmental considerations and the development of new bio-products in recent years, renewable sources as raw materials have grown and, consequently, are attractive both in industry and to the bioeconomy. The abundant availability of biomass as lignocellulosic materials from agricultural residues, forest, pulp and paper industries or from urban solid waste allows for its processing and different applications such as energy, fuels, chemicals, cosmetics, medical applications, construction materials and high added-value products for food or feed, which could be a commercial opportunity. For this reason, there is a clear opportunity to develop commercial processes that could generate products, at very high volumes and low selling prices, from biomass. It is important to consider that the development of efficient integrated processes and technologies for biomass conversion requires the effective utilization of all components. In general terms, and considering all the benefits, the production of high added-value compounds and bioenergy from biomass in the future will have an exponential behavior worldwide. High-pressure processing is a potential

Green Chemistry Series No. 48
High Pressure Technologies in Biomass Conversion
Edited by Rafał M. Łukasik

Published by the Royal Society of Chemistry, www.rsc.org

clean technology to convert biomass as raw materials into high added-value products and bioenergy. In this technology CO_2–H_2O mixtures, at high temperatures (160–250 °C) and pressures (above 200 bars), are applied for hydrolysis, extraction and structural modification of this biomass. Moreover, this process can be considered an ecofriendly technology, and in terms of a biorefinery, the high-pressure technology as a pretreatment stage plays an important role since the pretreatment allows the fractionation of the main components of the biomass producing high added-value compounds and substrates for conversion into bioenergy. Thus, the use of all products and co-products in the process of pre-treatment is essential, using the concept of biorefinery, which will help reduce the overall environmental and economic impact in the processes. This book discusses the tendency and developments concerning the use of high-pressure processing for the valorization of biomass. It reflects the current state of knowledge of high-pressure technology with an emphasis on fundamentals, chemical proprieties, phase diagrams, thermodynamic proprieties of CO_2, H_2O and the CO_2–H_2O mixture; the development of a sustainable technology in the biorefinery concept, the CO_2 for biomass pretreatment and in the biocatalysis process, the effect of high-pressure CO_2 and CO_2–H_2O mixtures on the direct conversion of biopolymers and as anti-solvent, production of high added-value compounds as furfural, 5-hydroxymethyl furfural (5-HMF), levulinic acid, valerolactone and the perspectives of high-pressure technology. This book will be a useful tool for scientists, engineers and researchers in both academia and industry. The conversion of biomass into high added-value compounds and bioenergy is essential to sustain our present and future and that high-pressure technology is one of the keys to achieving this goal using the biorefinery philosophy.

Preface

After the publication of a review entitled *Carbon Dioxide in Biomass Processing: Contributions to the Green Biorefinery Concept* published in Chemical Reviews[1] and the book *Ionic Liquids in the Biorefinery Concept*,[2] there was a need for a book that would reveal the details about the involvement of high-pressure technologies used in biomass processing. One of the first hurdles in this challenge was how to differentiate this book from dozens of other reviews, handbooks and university scripts dealing with the subject of high-pressure extraction of value-added compounds from biological matrixes. The book should not be another work about extraction. That is why this book is not about extraction, at least not about the extraction of volatiles from biomass matrix. The second hurdle was to establish frontiers in terms of the high-pressure processes considered in this book. One may suggest that steam explosion is a high-pressure technology, for example. Also, supercritical water processes of biomass gasification are also high-pressure technologies. Again, this book aims to avoid the description of these technologies as they are an industrial reality, also often obsolete technologies and are broadly presented in numerous books.

Hence there is a question: what does this book offer? This book aims to show the different aspects of high-pressure fluids in biomass processing. It starts with a general introduction to high-pressure technologies with special attention given to processes with water, as it is a fluid always present in biomass. Next, an introduction to the essential properties of high-pressure CO_2 and mixture with water and the benefits from the use of high-pressure fluids in biomass conversion are given. The third chapter tackles one of the most important aspect of biomass processing *i.e.* biomass pre-treatment. The role of high-pressure water and CO_2–H_2O mixture technologies is presented. This chapter is followed by one that describes a potential use of high-pressure fluids in enzymatic hydrolysis. Next, a series of chapters about the

Green Chemistry Series No. 48
High Pressure Technologies in Biomass Conversion
Edited by Rafał M. Łukasik

Published by the Royal Society of Chemistry, www.rsc.org

valorization of biomass fractions and the products of hydrolysis of these fractions are presented. The book finishes with perspectives about the development of high-pressure technologies in biomass processing.

The construction of this book allows the reader to dive in the subject of biomass processing with high-pressure fluids starting at a general level and moving on to a more specific and deeper analysis of each of the aspects of biomass processing from a high-pressure fluid perspective.

Additionally, due to the high-quality specialists authoring the chapters in this book, two relevant features have been integrated: (1) engineering aspects of high-pressure technologies and (2) a unique knowledge about complex biomass chemistry.

It is my pleasure to thank all the authors for their commitment and their highly valuable and professional contribution. I also wish to thank Fundação para a Ciência e a Tecnologia (Portugal) for their financial support.

Rafał M. Łukasik
National Laboratory of Energy and Geology, Lisbon, Portugal

References

1. A. R. C. Morais, A. M. da Costa Lopes and R. Bogel-Lukasik, *Chem. Rev.*, 2015, **115**, 3–27.
2. R. Bogel-Lukasik, *Ionic Liquids in the Biorefinery Concept*, RSC, Cambridge, UK, 2015.

Contents

Chapter 1 Supercritical Fluids in Natural Product and Biomass Processing – An Introduction 1
Manuel Nunes da Ponte

1.1 The Early History of Supercritical CO_2 Extraction 1
1.2 The Role of Water in Supercritical CO_2 Extraction 3
1.3 Fractionation of Liquids 4
1.4 Supercritical H_2O 5
1.5 Perspectives 6
References 7

Chapter 2 Introduction to High Pressure CO_2 and H_2O Technologies in Sustainable Biomass Processing 9
Ydna M. Questell-Santiago and Jeremy S. Luterbacher

2.1 Introduction 9
2.2 Biomass as Feedstock 11
2.2.1 First Generation Biofuels and Bioproducts – Edible Crops 11
2.2.2 Second and Third Generation of Biomass – Non-edible Crops 13
2.3 The Biorefinery Concept 14
2.3.1 Biorefinery Products 15
2.3.2 Main Biorefinery Processes 15

Green Chemistry Series No. 48
High Pressure Technologies in Biomass Conversion
Edited by Rafał M. Łukasik

Published by the Royal Society of Chemistry, www.rsc.org

2.4 High-pressure CO_2 and CO_2–H_2O Systems Within the Biorefinery Concept 19
2.4.1 Essential Features of High-pressure CO_2 and CO_2–H_2O Systems 19
2.4.2 Physical Processes Employing High-pressure CO_2 or CO_2–H_2O Systems 20
2.4.3 Chemical Processes Employing High-pressure CO_2 or CO_2–H_2O Systems 21
2.4.4 Challenges for Implementing Processes Using CO_2 30
2.5 Conclusion 30
References 31

Chapter 3 Pre-treatment of Biomass Using CO_2-based Methods 37
Luiz P. Ramos, Fayer M. De León Mayorga, Marcos H. L. Silveira, Célia M. A. Galvão and Marcos L. Corazza

3.1 Introduction 37
3.2 CO_2 Properties Under Subcritical and Supercritical Conditions 40
3.2.1 Physicochemical Properties and Phase Behaviour of CO_2 41
3.2.2 Dielectric Constant 47
3.2.3 Phase Equilibrium 49
3.3 Application of CO_2 for Biomass Pre-treatment and Fractionation 51
3.3.1 Use of $scCO_2$ Under Subcritical Water (CO_2–H_2O Mixtures) 51
3.3.2 Use of CO_2 Under Supercritical Conditions 54
3.4 Use of Co-solvents in CO_2-based Pre-treatment Methods 57
3.5 Scale-up of CO_2-based Methods for Biomass Pre-treatment 58
3.5.1 CO_2 Supply 59
3.5.2 CO_2 Pressurizing 59
3.5.3 Unit Operations 59
3.5.4 Reaction Feeding Mode 59
3.6 Conclusions 61
Acknowledgements 61
References 61

Chapter 4 Enzyme-based Biomass Catalyzed Reactions in Supercritical CO_2 66
Maja Leitgeb, Katja Vasić and Željko Knez

4.1 Introduction 66
4.2 Enzymatic Reactions in $scCO_2$ 67
4.2.1 Effects of Temperature and Pressure 70
4.2.2 pH of Medium and Formation of Carbonic Acid 71
4.2.3 Effect of Water Content 72
4.2.4 High-pressure Enzymatic Reactors 72
4.3 Biomass Conversion in $scCO_2$ 73
4.3.1 Algal Biomass in $scCO_2$ 77
4.4 Conclusion 78
Acknowledgements 79
References 79

Chapter 5 Direct Hydrolysis of Biomass Polymers using High-pressure CO_2 and CO_2–H_2O Mixtures 83
Ana Rita C. Morais and Rafal M. Lukasik

5.1 Introduction 83
5.1.1 Lignocellulosic Biomass Polymers 85
5.2 High-pressure CO_2 and CO_2–H_2O Mixture in the Hydrolysis of Biomass 88
5.2.1 Fundamentals 88
5.3 Hydrolysis of Biomass-derived Polymers 91
5.3.1 Cellulose 91
5.3.2 Hemicelluloses 93
5.3.3 Starch 98
5.3.4 Proteins 101
5.3.5 Lignin 102
5.4 Conclusions 110
Acknowledgements 110
References 110

Chapter 6 Processing of Lignocellulosic Biomass Derived Monomers using High-pressure CO_2 and CO_2–H_2O Mixtures 115
Gianluca Gallina, Pierdomenico Biasi, Cristian M. Piqueras and Juan García-Serna

6.1 Introduction 115

6.2 Cellulose and Hemicellulose Hydrolysis 117
6.2.1 The Phenomena at a Glance 117
6.2.2 Simple Mathematical Models to Describe Hydrolysis 118
6.2.3 The Main Reactions of the Monomers in Water: Tautomerization, Dehydration and Aldol Reactions 119
6.3 Reaction Medium and Operational Conditions 123
6.3.1 Subcritical Water and Carbonated Subcritical Water 124
6.3.2 Reactions in Supercritical Water 130
6.4 Reaction Configuration 132
6.5 Conclusions 133
References 134

Chapter 7 Efficient Transformation of Biomass-derived Compounds into Different Valuable Products: A "Green" Approach 137
Maya Chatterjee, Takayuki Ishizaka and Hajime Kawanami

7.1 Introduction 137
7.2 Experimental Methods 141
7.2.1 General Method for *in situ* Synthesis of Metal Nanoparticles Supported MCM-41 141
7.2.2 Catalyst Characterization 142
7.2.3 Catalytic Activity 142
7.2.4 Phase Behaviour Studies 145
7.3 Results and Discussion 145
7.3.1 Catalytic Strategies to Process 5-HMF into Fuel 145
7.3.2 5-HMF to 2,5-Dimethylfuran (DMF) 149
7.3.3 Furfural to 2-Methylfuran (2-MF) 155
7.3.4 THFA to 1,5-Pentanediol (1,5-PD) 156
7.4 Conclusion 160
Acknowledgements 161
References 161

Chapter 8 Anti-solvent Effect of High-pressure CO_2 in Natural Polymers **165**
Arturo Álvarez-Bautista and Ana Matias

8.1 Introduction 165
8.2 Biopolymers 166
8.2.1 Cellulose 166
8.2.2 Hemicellulose 167
8.2.3 Chitosan 168
8.3 Anti-solvent Effect 169
8.3.1 Anti-solvent Effect to Regenerate Biopolymers 169
8.3.2 Anti-solvent Effect of Compressed CO_2 171
8.3.3 Mechanisms of Precipitation 177
8.4 Perspectives 178
References 179

Chapter 9 Perspectives of the Development of High-pressure Technologies in Biomass Processing **181**
Rafal M. Lukasik

9.1 Perspectives 181
9.2 Conclusions 185
Acknowledgements 186
References 186

Subject Index **190**

CHAPTER 1

Supercritical Fluids in Natural Product and Biomass Processing – An Introduction

MANUEL NUNES DA PONTE

LAQV, REQUIMTE, Department of Chemistry, Faculdade de Ciências e Tecnologia, Universidade Nova de Lisboa, Portugal
Email: mnponte@fct.unl.pt

1.1 The Early History of Supercritical CO_2 Extraction

Thomas Andrews's 1869 Bakerian Lecture[1] "On the Continuity of the Gaseous and Liquid States of Matter" is widely credited to have established the term "critical point" to define the point in phase space where a liquid and its vapour attain the same density and become indistinguishable. In his lecture Andrews described, in detail, his experiments on carbonic acid (carbon dioxide). He stated that "On partially liquefying carbonic acid by pressure alone, and gradually raising at the same time the temperature to 88 °F, the surface of demarcation of the liquid and the gas became fainter, lost its curvature, and at last disappeared." Andrews goes on to establish the critical temperature of carbon dioxide as 30.92 °C, and to describe how, above this temperature there are no signs of phase separation, although the volume becomes extremely sensitive to pressure. He presents, in detail, the volume contractions obtained by small increases of pressure in the temperature region above the critical, up to 48.3 °C. He concludes that the volume exhibits a much greater contraction than it would if the perfect gas law was followed. He further concludes that, at higher pressures, the volume

Green Chemistry Series No. 48
High Pressure Technologies in Biomass Conversion
Edited by Rafał M. Łukasik

Published by the Royal Society of Chemistry, www.rsc.org

of the fluid takes similar values to those that might be calculated by using the thermal expansion of the liquid from below the critical temperature, that is, it behaves like a liquid.

This work inspired van der Waals to formulate his famous equation of state in his doctoral thesis "On the continuity of the gas- and liquid-state", presented in Leiden in 1873. This thesis had a profound effect in the development of molecular sciences in the late 19th century, ultimately leading to the award of the 1910 Nobel Prize for Physics to van der Waals.

A fluid in the pressure–volume–temperature region above the critical point where high compressibility and thermal expansivity were detected by Andrews is nowadays called a supercritical fluid. The high sensitivity of the density to small changes in pressure (or temperature) is the most distinctive property of supercritical fluids, and it forms the basis for their technological applications developed in the last forty years. It took, however, about a hundred years for Andrews's work on carbon dioxide to be translated into the industrial process known these days as supercritical fluid extraction.

Truly, through the years, processes like the high-pressure polymerization of ethylene, discovered by ICI in the 1930s, and the so-called ROSE (Residuum Oil Supercritical Extraction) process, developed in the petrochemical industry and using a light paraffinic solvent, like pentane, in supercritical conditions, can be counted as using supercritical fluid solvents. The massive work of Francis on mixtures of liquid carbon dioxide with hundreds of compounds, published in 1954 in one single paper,[2] certainly contains much relevant information in conditions that may be deemed "near critical". But it was not until Zosel,[3] working in the 1970s in the Max Planck Institute in Mulheim, Germany, used carbon dioxide to extract caffeine from coffee, and the technique went commercial, with an industrial plant inaugurated in 1978 by Kaffee HAG, that the field was really launched. The first symposium dedicated to the subject was held in Essen, Germany, in the same year, and influential books started to appear at the beginning of the 1980s, by Schneider, Wilke and Stahl[4] and by McHugh and Krukonis.[5] The development of the field was rapid, as it attracted researchers from many different areas. Ten years after the Essen Symposium, in 1988, a totally dedicated journal (*The Journal of Supercritical Fluids*) appeared, and the first International Symposium on Supercritical Fluids was held in Nice, France, with widespread international participation.

To a large extent most investigators were drawn to the area by the credentials of carbon dioxide as an environmentally safe solvent that could replace volatile organic solvents. This was the case with the first above-mentioned application, decaffeination, and in the second large scale commercial use, the extraction of hops for the beer industry. These days, most large-scale applications remain in natural products. One of the last processes to attain commercial status uses extraction volumes of the size (or bigger) of coffee decaffeination, and it cleans cork in the Spanish plant of the company Diam, described by Lack.[6] The plant has recently undergone a duplication of capacity and the company is building a new plant in France, which might turn this process into the largest one in terms of the overall volume of the installed extractors.

Supercritical CO_2 extraction has been thoroughly reviewed and explained in the 1994 book of Brunner.[7] At about the same time, the second edition of the book of McHugh and Krukonis[8] was published and a book edited by King and Bott[9] described in detail extraction processes like caffeine from coffee or hops for the beer industry. With the appearance of these books, it may be said that the field attained full maturity and entered the array of well-known separation operations that can be applied to natural products and biomass.

1.2 The Role of Water in Supercritical CO_2 Extraction

Extraction processes use carbon dioxide in essentially two sets of conditions: at high density (higher pressure), where it can dissolve the target solutes, and at low density (lower pressure), where it precipitates the solutes. The high compressibility of carbon dioxide at supercritical conditions allows its solvent power to be rapidly varied with relatively small changes in pressure (and with temperature, but changes in temperature are usually more difficult to implement).

Carbon dioxide is itself a bad solvent. As a small molecule with little polarity (no permanent dipole moment, a small quadrupole), it dissolves mainly non-polar molecules of small molar mass. This is exactly what is required, for instance, in the case of the cleaning of cork, where one contaminant must be removed, but most natural constituents of cork should be left untouched to maintain the properties that make it the undisputed leader of wine bottle closure materials. Low carbon dioxide pressures, of the order of 10 MPa, are therefore used in this process. However, in the extraction of caffeine, high pressures (30 MPa) and high solvent power are sought, due to the low solubility of this target substance in carbon dioxide. In this case, coffee beans are extracted before roasting, so that the aroma constituents, so characteristic of coffee, are not present yet (and therefore are not extracted).

In both cases,[10,11] the presence of water in the biomass extraction cake is absolutely fundamental for the extraction to proceed. This requirement is "often misunderstood" – a quote from McHugh and Krukonis (ref. 8, p. 294). These authors explain in detail that in all cases of extraction of caffeine from coffee, and not only with carbon dioxide, moisture is essential to "free caffeine from its bound state in the coffee matrix". Caffeine is soluble in dry carbon dioxide, but it will not be available for extraction from the coffee matrix without the presence of water.

As all biomass contains water, these effects are in fact quite general in the extraction of natural products. For instance, Brunner,[7] discusses at length the effect of the content of water in cocoa seed shells on the yields of carbon dioxide extraction of theobromine.

But, as pointed out by McHugh and Krukonis, those effects are often misunderstood and confounded with the role played by entrainers, modifiers or co-solvents, substances that are added to carbon dioxide to increase the solvent power of the supercritical solvent towards bigger or more polar molecules. Ethanol and propane have been the most researched co-solvents, but other small molecules, like methanol or acetone, have also been studied. One required property of these substances is that they must be mutually soluble with carbon dioxide in all proportions, at the pressures and temperatures of extraction, so that their concentration in the solvent may be controlled at the inlet of the extraction vessel and remain constant throughout the process. Typical concentrations used are 5 or 10% in mass.

Quite differently, water is scarcely soluble in supercritical carbon dioxide. Its solubility, in the range of pressures and temperatures commonly used in supercritical CO_2 extraction, is of the order of 2 g per kg of carbon dioxide. It cannot therefore much enhance the solvent power of the gas. Dry carbon dioxide circulating in a moist plant matrix will soon become saturated in water. If water is present in large quantities in the biomass in the extractors, it can be dragged as a liquid, and eventually freeze in the decompression area, leading to tube blockages and other operational problems. The water content of the plant material to be subject to extraction must therefore be carefully balanced.

Even in the case where dried raw-materials are used, the water content of circulating carbon dioxide will be approximately constant. Reverchon *et al.*[12] in their studies of the extraction of dried sage leaves measured a constant concentration of water in the carbon dioxide leaving the extractor of about 20% of the saturation value. They reasoned that, in the dried material, there were no free water molecules to be taken by the solvent, but many remained bound to the solid material. The constant value they measured would result from an adsorption–desorption equilibrium.

In recent times, the scientific interest on carbon dioxide + water mixtures was renewed,[13] due to the considerable efforts currently developed in the study of *Carbon Capture and Sequestration* (CCS), a global warming mitigation scheme. The main interest is driven by the possibility that highly salted water reservoirs deep in the Earth may act as carbon dioxide sinks, and be useful for sequestration. Curiously, if these methods do live up to expectations and CCS is implemented, the enormous scale of the carbon dioxide that will be made available will certainly contribute to a renewal of the field of supercritical CO_2.

1.3 Fractionation of Liquids

Supercritical CO_2 fractionation of liquid mixtures has been intensively studied in the first years of the development of the supercritical fluid area. Contrarily to extraction from solids, which has benefited from the visibility given by reasonably large-scale applications (caffeine, hops), the fractionation of liquids still lacks a flag-carrying well-known commercial process to establish itself. In this process, tall and slim counter current columns are used to increase the number of separation stages. It has the advantage that it may be processed continuously, as in a liquid–liquid separation, all feed materials being liquid.

As carbon dioxide is generally a weak solvent, this type of fractionation may be used for difficult separations in liquid mixtures. It has been proposed for the fractionation of valuable substances in natural products, like vegetable food oils, fish oils, wine and beer. Commercial applications are reported for small-scale productions, as in the case of the purification of omega 3 fatty acids (see, for instance, www.solutex.es). Brunner[7] describes the fundamentals and design techniques in great detail and Brunner and Machado[14] have more recently provided a good example of the application of fatty acid fractionation, while providing a thorough recollection of previous work in supercritical CO_2 fractionation.

Very recently, Bejarano *et al.*[15] have thoroughly reviewed the field. In their paper, they present a comprehensive list of existing facilities throughout the world and describe, in detail, the different types of fractionation columns used so far, as well as the data needed to design and operate them.

1.4 Supercritical H_2O

One major event in the supercritical fluid field was the appearance of supercritical water. It developed as a second important focus of detailed study, so much so that the whole field is now supported on two pillars, supercritical CO_2 and supercritical H_2O, with little activity dedicated to other substances. Water has, however, a high critical temperature (374 °C) and pressure (22 MPa), and working with it involves much harsher conditions than with supercritical CO_2. Moreover, the changes induced by temperature increases on liquid water solvent properties are much more drastic than for liquid carbon dioxide. In fact, the three-dimensional hydrogen bond network, characteristic of water at room temperatures (responsible for its high solvent power towards ionic salts) is gradually disturbed by molecular agitation as the temperature rises. It is almost completely disrupted at temperatures well below the critical, where water starts to behave as a highly polar, but, paradoxically, "non-aqueous" solvent. Supercritical water can dissolve hydrocarbons, a positive property, but precipitates ionic salts, which constitutes one of the main problems for its devised applications. This evolution of the solvent power of water with temperature may be summarized by its dielectric constant, which changes from around 80 at room temperature and pressure to about 6 just above the critical conditions.

Another property that also changes drastically with temperature is the ionic product K_w, which increases by several orders of magnitude from room to critical temperature. This means that hydrolysis reactions are highly facilitated in supercritical H_2O, as pointed out by Weingartner and Franck.[16] The corrosion of piping and reactor surfaces has therefore represented an important drawback in the development of supercritical H_2O applications.

One of the most studied processes has been supercritical H_2O oxidation (SCWO), mainly focused on the destruction of organic residues. Water (in supercritical conditions) mixes in all proportions with oxygen, and complete oxidation of carbon into carbon dioxide and hydrogen into water can be very rapidly obtained at the highly reactive conditions of supercritical H_2O. Toxic by-products, such as dioxins, of more classical incineration processes are thus avoided. However, when other elements are present in the residues, such as chlorine, sulfur or phosphorus, acids (hydrochloric, sulfuric, phosphoric) or the corresponding salts are formed, this leads to the two most important problems of SCWO: corrosion and plugging by salt deposition.

Marrone[17] has recently reviewed the commercial activity in this area. He thoroughly lists the companies active in the market and the facilities currently in operation, as well as those that, after some period of activity, have been closed. It is very interesting to note the high "mortality" rate of both companies and facilities, mostly because operations were plagued by technical problems and could not attain specifications. But it is also pointed out that new companies are entering the market and that SCWO continues to be an attractive technology.

Supercritical H_2O has been used as a medium for many interesting reactions, especially to produce materials. Adschiri *et al.*[18] reviewed the activity at an industrial scale in Japan and Korea. In the handling of biomass, supercritical and hot subcritical water have been proposed[19] as advantageous media to produce biofuels. These highly reactive "hydrothermal media" promote the depolymerisation of the main biomass constituents, leading to the production of either liquid or gaseous fuels, depending on operational conditions and biomass composition.

1.5 Perspectives

Forty years on from the initial thrust focusing on supercritical CO_2 and extraction of natural products, the field has diversified into a wide range of areas of study and many applications, which either have already gone commercial or are waiting to be adopted. A rough measure of this diversity may be given by the 2009 special edition issue of *The Journal of Supercritical Fluids*, commemorating its 20th anniversary.[20] 32 review papers were published, covering the whole field. They may be loosely divided in the following topics: supercritical H_2O – 8; processing of polymers, mostly with carbon dioxide – 6; other materials – 2; separations (extraction, fractionation, chromatography) with supercritical CO_2 – 6; colloids, microemulsions and particle formation – 5; reaction in supercritical CO_2 – 4; and energy applications – 1.

Although the role of natural products and biomass is no longer the dominant focus of the field, it continues to play an important part in recent developments. The properties of supercritical fluids as green solvents are still an attractive feature, as seen in the recent review of Farrán *et al.*[21] on green solvents in carbohydrate chemistry.

The combination of carbon dioxide with hydrothermal technologies, as those proposed by Brunner and collaborators[22] for ethanol production or King[23] for multiple separations from plant material, represent technological platforms that can be used in a decentralized manner for sustainable use of biomass. The inspirational view of Arai, Smith and Aida[24] on sustainability and supercritical H_2O processes may, in reality, be extended to the whole field.

Curiously, as pointed out above, some of the technologies under study for mitigation of the consequences of the (unsustainable) burning of fossil fuels may provide soon the materials needed for much increased applications of supercritical fluids. On one hand, as pointed out by Brunner in his recent review of supercritical fluids and energy,[25] the steam cycles used in power plants to recover heat from flue gases have been directed more and more towards supercritical conditions, due to the higher thermodynamic efficiencies obtainable with higher hot reservoir temperatures. This trend will contribute to the dissemination of supercritical water technology and the improvement of the materials used in piping and reactors. On the other hand, large quantities of carbon dioxide may start to become available for utilisation by the implementation of carbon capture techniques in power plants. The area of carbon dioxide utilisation is growing fast in academia and is attracting attention both from industry and from regulators, as an alternative to carbon sequestration. A promising future for supercritical fluids and their applications seems to be waiting.

References

1. T. Andrews, *Philos. Trans. R. Soc. London*, 1869, **18**, 42–45.
2. A. W. Francis, *J. Phys. Chem.*, 1954, **58**, 1099–1114.
3. K. Zosel, *Angew. Chem., Int. Ed.*, 1978, **17**, 702–709.
4. G. M. Schneider, G. Wilke and E. Stahl, *Extraction with Supercritical Gases*, Vch Pub, 1980.
5. M. McHugh and V. Krukonis, *Supercritical Fluid Extraction–Principles and Applications*, Butterworth-Heineman, Boston, MA, USA, 1986.
6. E. Lack and H. Seidlitz, Industrial cleaning of cork with supercritical CO_2, in 3rd International Meeting on High Pressure Chemical Engineering, Erlangen, Germany, 2006.
7. G. Brunner, *Gas Extraction*, Steinkopff Verlag, 1994.
8. M. McHugh and V. Krukonis, *Supercritical Fluid Extraction–Principles and Applications*, Butterworth-Heineman, Boston, MA, USA, 1994.
9. M. King and T. R. Bott, *Extraction of Natural Products Using Near-critical Solvents*, Springer Science & Business Media, 2012.
10. D. Chouchi, C. Maricato, M. Nunes da Ponte, A. Pires and V. San Romao, SFE of trichloroanisole from cork, in Proceedings of the International Symposion on Supercritical Fluids, Sendai, Japan, 1997.
11. E. Lack and H. Seidlitz, in *Extraction of Natural Products Using Near-Critical Solvents*, Springer, 1993, pp. 101–139.

12. E. Reverchon, R. Taddeo and G. D. Porta, *J. Supercrit. Fluid.*, 1995, **8**, 302–309.
13. S.-X. Hou, G. C. Maitland and J. M. Trusler, *J. Supercrit. Fluid.*, 2013, **73**, 87–96.
14. G. Brunner and N. Machado, *J. Supercrit. Fluid.*, 2012, **66**, 96–110.
15. A. Bejarano, P. C. Simões and J. M. del Valle, *J. Supercrit. Fluid.*, 2016, **107**, 321–348.
16. H. Weingartner and E. U. Franck, *Angew. Chem., Int. Ed.*, 2005, **44**, 2672–2692.
17. P. A. Marrone, *J. Supercrit. Fluid.*, 2013, **79**, 283–288.
18. T. Adschiri, Y.-W. Lee, M. Goto and S. Takami, *Green Chem.*, 2011, **13**, 1380–1390.
19. A. A. Peterson, F. Vogel, R. P. Lachance, M. Froling, M. J. Antal and J. W. Tester, *Energ. Environ. Sci.*, 2008, **1**, 32–65.
20. E. Kiran, G. Brunner and R. L. Smith Jr., *J. Supercrit. Fluid.*, 2009, **47**, 333–636.
21. A. Farrán, C. Cai, M. Sandoval, Y. Xu, J. Liu, M. J. Hernáiz and R. J. Linhardt, *Chem. Rev.*, 2015, **115**, 6811–6853.
22. C. Schacht, C. Zetzl and G. Brunner, *J. Supercrit. Fluid.*, 2008, **46**, 299–321.
23. J. W. King and K. Srinivas, *J. Supercrit. Fluid.*, 2009, **47**, 598–610.
24. K. Arai, R. L. Smith and T. M. Aida, *J. Supercrit. Fluid.*, 2009, **47**, 628–636.
25. G. Brunner, *J. Supercrit. Fluid.*, 2015, **96**, 11–20.

CHAPTER 2

Introduction to High Pressure CO_2 and H_2O Technologies in Sustainable Biomass Processing

YDNA M. QUESTELL-SANTIAGO AND
JEREMY S. LUTERBACHER*

Laboratory of Sustainable and Catalytic Processing, Institute of Chemical Sciences and Engineering, École Polytechnique Fédérale de Lausanne (EPFL), Station 6, CH.H2.545, Lausanne 1015, Switzerland
*Email: jeremy.luterbacher@epfl.ch

2.1 Introduction

Petroleum is a major raw material for the production of fuels, chemicals and materials used in our daily lives. The continuous growth in the consumption of these products requires the increased exploitation of fossil resources. In turn, this exploitation has led to environmental and economic issues linked to climate change, resource depletion and political instabilities due to the unequal distribution of fossil deposits.[1] The global primary energy consumption increased by a constant rate of 2.4% per year $\pm$0.08% since 1850 and shows no sign of slowing down.[2] The global community, increasingly aware of the dangers linked to diminishing fossil reserves and climate change, is encouraging the development of renewable carbon sources.

Green Chemistry Series No. 48
High Pressure Technologies in Biomass Conversion
Edited by Rafał M. Łukasik

Published by the Royal Society of Chemistry, www.rsc.org

In 2013 the world total energy supply was 13.5 Gtoe and only 10.2% of this energy was supplied by biofuels and wastes.[3] According to the International Energy Agency (IEA) this fraction is projected to increase to 27% of the energy demand in the transportation sector by 2050.[4]

Using biomass and organic wastes as a feedstock for renewable carbon represents a promising and sustainable alternative to fossil carbon.[5] However, exploiting biomass fully is associated with many challenges including the development of suitable conversion technologies both from an environmental and economic standpoint coupled with political support issues. The production of fuels from biomass has been criticized due to the so-called fuel *vs.* food competition. The potential for large-scale biomass exploitation to drive up the price of competing food crops has led to increased attention to the substitution of non-fuel petroleum-based materials and jet fuel by biomass-derived molecules. Both jet fuels and these carbon-based chemicals are difficult or impossible to substitute with renewable electricity and are potentially more valuable than other transportation fuels. Furthermore, the demand for non-fuel carbon-based molecules and jet fuel is significantly lower than for total fuel and therefore leads to less competition issues for land used in food production. For these reasons, the production of jet fuel and carbon-based chemicals from biomass has attracted less controversy and has helped promote the concept of an integrated biorefinery.[6] This concept is defined as the production of at least one energy product (besides heat and electricity) and the production of at least one high value chemical or material, along with low-grade and high-volume products in a single biomass conversion plant.[7]

Biorefineries include multiple conversion technologies used in a sustainable manner to comply with the production of the diverse biomass feeds mentioned. Current conversion technologies are usually designed for a specific type of biomass and can suffer from low yields, high energy requirements and elevated operation costs that makes the integration of bioproducts difficult in a competitive market. Several approaches have attempted to improve the efficiency of biomass processing while applying sustainable development principles. Here, we provide an overview of the use of CO_2 and CO_2–H_2O mixtures for biomass processing and how such approaches have contributed to the development of more efficient and sustainable processes. We will begin by discussing the characteristics and categories of biomass feedstocks and will briefly describe the typical processes employed in biorefineries with their main bioproducts. We will then introduce the principle applications of CO_2 and CO_2–H_2O mixtures in biomass processing by reviewing the main features of this system. We also provide an overview of the principle applications where CO_2 and CO_2–H_2O mixtures are used and discuss the advantage that their use has provided. Finally, we briefly examine the challenges for implementing processes using high pressure CO_2.

2.2 Biomass as Feedstock

Biomass refers to a renewable organic material created biologically and is defined as "a substance wholly comprised of living or recently living (non-fossil) material".[8] This concept includes an extensive range of materials, *e.g.* whole or parts of, plants, trees, animals, microorganisms and aquatic organisms, which are classified as edible and non-edible crops. When considered as a source of fuel, biomass is sub-classified into three categories referred to as first, second or third generation biofuels.[8,9] First generation biofuels or bioproducts refer to fuels and chemicals made from edible crops that are used for the production of products containing starch (*e.g.* corn, flower, *etc.*), sugar, vegetable oil, lipids and/or proteins. Second generation biofuels or bioproducts refer to molecules made from lignocellulosic materials, which are non-edible crops, mainly composed by polysaccharides. Lignocellulosic material tends to be more difficult to convert to fuels but usually provides more material per plant growth area and requires less fertilizer to grow.[10,11] Both of these attributes lead most experts to agree that second generation biofuels and bioproducts are more sustainable than their first generation equivalents. The composition of lignocellulosic material is usually about 40–50% cellulose, 25–35% hemicellulose and most of the remainders is lignin.[1,12] Products produced from microalgae have sometimes been referred to as third generation biofuels and bioproducts. Algae has mainly been considered for the production of oil and lipids.[9,13] Figure 2.1 shows the chemical structure of the main components of the feedstocks used for the three generations of bio-based molecules.

As discussed, biomass has the potential to replace fossil-based feedstock for the production of chemical intermediates and fuels. However, the chemical structure and composition of biomass is considerably different from that of crude oil, which is the major conventional fossil feedstock.[14] Hence, all biomass processing employ reforming steps to produce direct or indirect petrochemical replacements. We briefly discuss the principal reforming efforts for each generation of bio-derived products below.

2.2.1 First Generation Biofuels and Bioproducts – Edible Crops

First generation biofuels and bioproducts are produced from biomass with significant edible fractions, such as corn, sugar cane, wheat, palm oil and rapeseed. These feedstocks are the primary sources of starch, sugars and vegetable oil, mainly used to produce energy products that include biogas, bioethanol and biodiesel.[7] Biogas is produced by the anaerobic digestion of mixtures of starch, manure and other organic wastes, and is used to produce electricity and in some countries as transportation fuel, after purification and pressurization of the resulting methane. In such cases, starch, free sugars, oils and proteins can generally be converted to gas by microorganisms

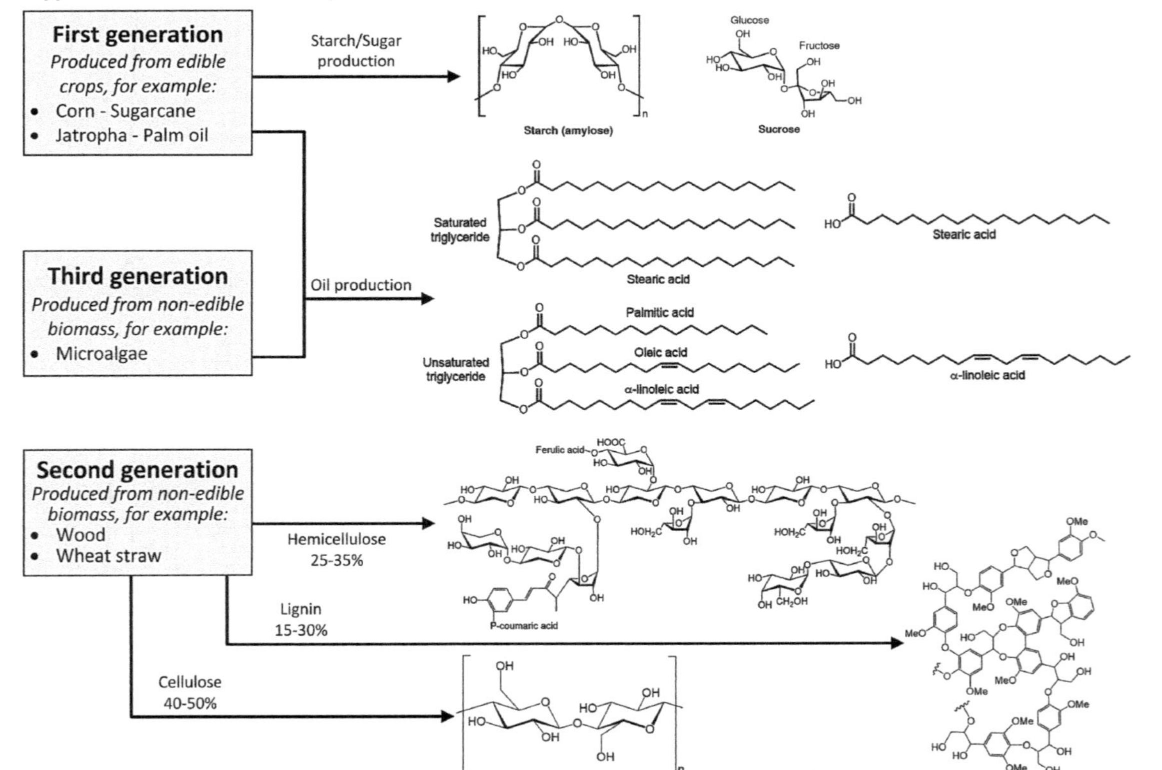

Figure 2.1 The main components of all three major biomass categories.

but other inedible biomass fractions generally remain untouched. Bioethanol is recovered from the fermentation of simple sugars derived directly from plants such as sugarcane, sugar beets or indirectly from starch-containing crops. Simple sugars are fermented directly, while amylase enzymes are generally added to the starch-based biomass before fermentation to break down the starch to simple sugars. Currently, bioethanol is the largest biofuel produced worldwide with a total world production of 94 billion litres of bioethanol in 2014.[15] The largest producer of bioethanol in 2014 was the USA with a production of 54.5 billion litres, Brazil was second with 26.3 billion litres and Europe third with 5.6 billion litres. Another transportation fuel with a large worldwide production capacity is biodiesel, which is produced from oil-based crops including palm oil, rapeseed and soybean. The two world leaders in biodiesel production in 2014 were Europe (11.7 billion litres) and the USA (4.8 billion litres).[15]

Feedstocks used to produce first generation biofuels and bioproducts are very easy to convert to biofuels due to their availability within the plant structure (which is why these fractions are considered edible). However, first generation products have several issues including their competition with food, their limited availability, their high fertilizer use and, therefore, their minimal savings in terms of CO_2 emissions compared to their fossil equivalent. For these reasons, the global biofuel demand in 2050 is not expected to be met by edible crops and is rather expected to be supplied almost entirely from second and third generation biofuels and bioproducts, which represents a future requirement of *ca.* 30 ExaJoules (EJ).[16]

2.2.2 Second and Third Generation of Biomass – Non-edible Crops

Lignocellulosic biomass, microalgae and organic wastes are described as non-edible crops. The bulk of terrestrial biomass is represented by lignocellulosic materials, composed mainly by three natural polymers: cellulose, hemicellulose and lignin. Agriculture and forestry residues are examples of lignocellulosic biomass that are being considered for the production of a wide variety of biofuels and fine chemicals. Common upgrading routes often involve the fractionation and depolymerization of hemicellulose, cellulose and lignin to produce 5-carbon sugars, 6-carbon sugars and aromatic chemicals, respectively. However, other direct processes target platform molecules such as dehydration products or sugar hydrogenation to alcohols.[1] Organic wastes mainly refer to sewage sludge, pulp and paper mill sludge, food waste, manure and other agricultural residues, which are composed of carbohydrates, proteins, lipids and lactose.[17,18] These feedstocks can be directly processed or pre-treated prior to conversion for the production of biogas, hydrogen and C_2 to C_4 hydrocarbons.[18] Microalgae (sometimes referred to as the feedstock for third generation biofuels and bioproducts) are being considered mainly for the production of biodiesel from extracted lipids by catalytic or enzymatic transesterification.[19] Due to their higher photosynthetic efficiency, they have an impressive potential for the production of biofuels in comparison to lignocellulosic biomass, although this potential has never been achieved at an industrial scale.[20] Sugarcane production can reach values around 4900 kg of bioethanol per hectare,[21] while microalgae's potential has been said to reach 52 000–121 100 kg of biodiesel per hectare.[20] Microalgae can also be used as a source of natural dyes, antioxidants, carbohydrates and other fine chemicals.[20] After extraction, the remaining algae can be processed into ethanol, methane, livestock feed or fertilizer.

2.3 The Biorefinery Concept

A biorefinery is a facility where a sustainable process integrates the production of fine chemicals, materials, biofuels and heat/power from biomass with the least amount of leftovers after treatment. The representation of an integrated biorefinery incorporating the re-utilization of CO_2 within its processes is shown in Figure 2.2. The concept of a biorefinery is analogues to today's petroleum refineries, where multiple carbon-based products and fuels are produced from crude oil.[7] Proposed biorefinery processes employ a wide range of technologies to fractionate biomass into valuable compounds (polysaccharides, sugars, oils, lipids, proteins) and for subsequent upgrading these compounds in subsequent steps. Feedstock fractionation is a crucial step due to the heterogeneity of the biomass and the multiple functional groups that are present.

Currently, most bio-products produced from edible crops are manufactured in single production chains and not within a biorefinery. Because these plants have already been built, so far, the main focus of processing plants based on edible crops has been to further optimize their processes and reduce costs rather than implement new technologies.[12] However,

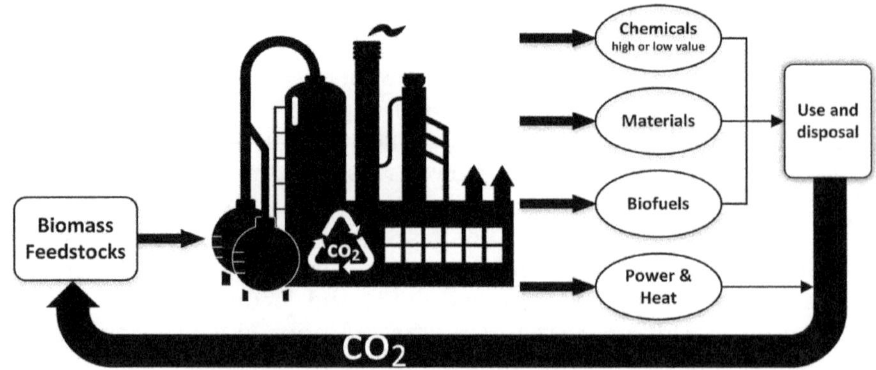

Figure 2.2 A schematic representation of the biorefinery concept with integrated CO_2 use and recycling.

biorefineries based on second and third generation biofuels and bioproducts have the opportunity to exploit many more routes and achieve larger scales due to a larger biomass potential yield per cultivation area, which limit biomass transportation restrictions.[7] Such scales could also allow the implementation of new conversion technologies that only become economical at very large scales, which include the high-pressure systems required for high-pressure CO_2 use.

2.3.1 Biorefinery Products

Presently, the chemical industry refines crude oil into fractions that include naphtha (from which all the major bulk chemicals are derived), gasoline, kerosene, gas oil and residues.[22] The processes employed in the refinery industry include numerous cracking and refining catalysts as well as distillation as the dominant separation strategy. An important characteristic and advantage of the naphtha fraction compared to biomass, is its low oxygen content. Most bulk chemicals produced in refineries are derived from molecules containing no oxygen such as ethylene, propylene, C_4-olefins and the aromatics benzene, toluene and xylene (BTX).[22]

In principle, most petrochemical refinery platform molecules could be derived from renewable carbon sources. Unfortunately, this is currently only possible at relatively low yields and high costs.[22] For this reason, it has been proposed that the biorefinery industry produce a number of petrochemical substitutes through a selection of simple platform molecules that are different from those currently used in the petrochemical industry.[7] Given the chemical complexity of biomass, there is a range of platform chemicals that could be produced from one type of feedstock depending on the chosen processing strategy. Nevertheless, several of these building block chemicals are expected to be derived from the carbohydrate fraction of biomass, which, due to it being a major component of all types of biomass, is likely to play a crucial role in future biorefineries. In 2004, the US Energy Department identified twelve building block chemicals of major importance that could be produced from sugar by biological or chemical conversions.[23] These twelve sugar-based building block chemicals are shown in Figure 2.3 along with the main chemicals and products (in bold) that can be easily derived from these intermediates and that are likely to be found in an integrated biorefinery.

2.3.2 Main Biorefinery Processes

The main processes in proposed biorefineries will likely be used to fractionate, depolymerize and deoxygenate biomass components. Since biomass is already highly oxidized, several hydrogenation and/or dehydration transformations are usually required. These transformations are key steps for the conversion of biomass into building blocks molecules and value-added compounds. These processes can be classified into two main categories

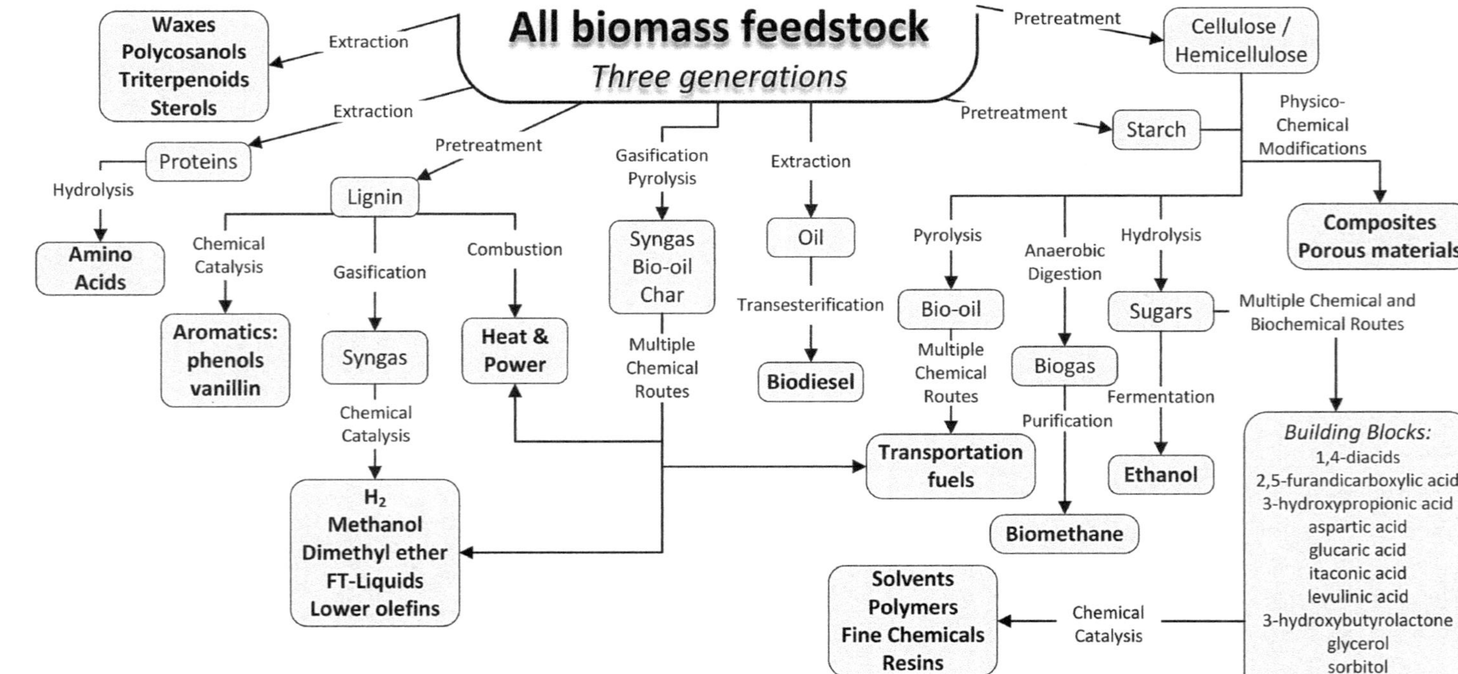

Figure 2.3 The principal conversion routes and products found in an integrated biorefinery.

depending on the nature of biomass transformation: physical or chemical treatments.

Physical or mechanical processes are usually involved as the first step in biorefineries as a preconditioning step, although they are not always required.[7] These processes do not change the composition of the biomass, but only lead to particle size reduction or separation of the feedstock components. Biomass size reduction refers to changes in particle size, shape or bulk density within specific ranges depending on further conversion processes requirements. Separation procedures can consist of increasing the concentration of valuable compounds without the transformation of their components by extractions methods, mainly using organic solvents or supercritical fluids.

Chemical processes involve changes in the chemical structure of one or more of the molecules by introducing high temperatures or catalytic species into the reaction media. For the purpose of this chapter, we classified these processes as thermochemical and catalytic. We define thermochemical pathways as pathways that are rapid and largely driven by high temperatures rather than just by the presence of catalysts. These thermochemical processes have been used for the production of syngas, bio-oil, bio-char/bio-coal and power from biomass[24] and include gasification, pyrolysis and dry/wet torrefaction. Gasification consists of keeping biomass at high temperatures (>700 °C) with limited oxygen concentration to produce hydrogen, methane, syngas, and/or power. Hydrogen and methane can also be in liquid conditions at high pressure if hydrothermal gasification (HTG) conditions are used. HTG operates within the subcritical or supercritical region of water (374 °C and 22.1 MPa) usually with a catalyst and generally uses biomass feedstocks with a high-water content, *e.g.* manure, that benefit most from avoiding a pre-drying step in the process.[18,25] Syngas is an important intermediate in several biorefinery processes that could lead to multiple fuels or chemicals.[23] The Fisher–Tropsch (FT) process is a well-known route for the upgrading of syngas into long chain liquid hydrocarbon and alkanes and short alcohols (C_1 to C_3) by multiple catalytic syntheses using mostly cobalt and iron-based catalysts.[26] Other catalysts such as nickel and ruthenium, have shown high catalytic activity in FT processes, but under operational conditions nickel promotes methanation, which is undesirable for this route, and ruthenium is expensive, which increases process costs. Higher alcohols (C_4 to C_{10}), used as additives in the reformulation of gasoline[27] and for the production of C_2 to C_4 olefins,[28] are also produced from syngas using a several catalytic routes.

Pyrolysis is defined as the processing of biomass at temperatures between 300 °C and 600 °C in the absence of oxygen, to produce bio-oil, charcoal and light gases similar to syngas. Bio-oil is usually the most desirable product and its production can be maximized by applying what is known as fast pyrolysis conditions (which involves rapid heating up to 500 °C and short residence times at high temperatures, *e.g.* seconds).[24] Bio-oil can be upgraded to mixtures of molecules that can be used as transportation fuels or chemicals. However, the high production of char and coke, and the important hydrogen consumption still remain problematic.[7] Dry torrefaction, sometimes referred to as a mild pyrolysis, is performed under an inert environment at atmospheric pressure with a retention time around one hour at low temperatures (200 °C to 300 °C or 180 °C to 260 °C in case of wet torrefaction (WT)).[29,30] In both conditions the main product is bio-char or bio-coal, which is used as a feedstock for combustion or gasification. The advantage of this char over raw biomass is that its characteristics are more stable and that it has a higher energy density, which makes its transportation less environmentally and economically intensive.

Catalyzed chemical reactions play a vital role in the upgrading of building platform molecules to petrochemical equivalents and they can be found at various stages of biomass conversion. The most common catalyzed reactions that occur in biorefineries are hydrolysis, dehydration, hydrogenation and transesterification reactions. Hydrolysis uses acids, alkalis or enzyme catalysts to depolymerize polysaccharides and proteins to their corresponding sugars or amino acids. Acid catalyzed dehydration reactions and metal-catalyzed hydrogenation reactions are important deoxygenation reactions notably for sugars. Transesterification is used to produce biodiesel from lipids or oils by an acid or base-catalyzed reaction with a short alcohol such as methanol. Other important chemical reactions are FT synthesis, methanation and steam reforming, which involve syngas conversion in the gas phase and are used to produce hydrocarbons, methane or hydrogen, respectively.[7] Of all the biochemical reactions currently implemented in various industries, there are mainly three types found in biorefineries: fermentation, anaerobic digestion and enzymatic reactions. Fermentation and anaerobic digestion involve live cells whereas enzymatic-based reactions usually involve cell free mixtures of proteins. Fermentation generally refers to the production of specific non-volatile chemicals such as alcohols, organic acids or even alkanes from carbohydrates. CO_2 is almost always a co-product of fermentative processes. Anaerobic digestion refers to a subset of fermentation reaction that produce gaseous products (mainly methane and CO_2) from carbohydrate and proteins found in biomass and organic wastes. Enzymes are usually used to facilitate the deconstruction of various polymers including polysaccharides and proteins.

Each of these processes presents significant conversion challenges, which continues to encourage innovation within biorefinery research. Multi-disciplinary efforts have been carried out in the design of novel catalysts, high-pressure systems, solvent engineering and multi-scale modelling in order to surpass current process limitations.[14] Research in high-pressure CO_2 and CO_2–H_2O systems is an example of these efforts. In the past 20 years, research publications related to the application of CO_2 in green processes and biomass conversion has increased significantly.[31] Below, we outline the features of CO_2 and CO_2–H_2O systems that have contributed to this increased interest and the potential advantages of their use in biomass conversion.

2.4 High-pressure CO_2 and CO_2–H_2O Systems Within the Biorefinery Concept

2.4.1 Essential Features of High-pressure CO_2 and CO_2–H_2O Systems

Carbon dioxide is an attractive replacement for organic solvents because it is abundant, inexpensive, non-flammable, environmentally benign and can be easily recycled and disposed of.[32] Its physical properties such as small molecular size and absence of surface tension (at high pressure) grant CO_2 high penetrating capabilities and lead to high mass and heat transfer coefficients.[31] Despite its inertness, CO_2 can also participate as a catalyst or inhibitor, and interact with other catalysts and substrates, which can affect reaction performance. Presently, high-pressure CO_2 is often considered for high-pressure sustainable process development due in part to its low critical pressure and temperature,[9] which are easily reachable in industrial reactors.[31] The special combination of gas-like viscosity and diffusivity, and liquid-like density of high-pressure CO_2 makes it a highly tunable solvent and leads to many opportunities for process development. Moreover, when CO_2 is used in confined processes it can be considered as sequestered given that it was generally obtained from natural deposits or man-made production, *e.g.* ammonia plants.[31,33] Though opportunities offered by CO_2 use exist across many fields they extend to several processes that are being considered within biorefineries.

High-pressure CO_2 has a dielectric constant that depends on its density but always remains relatively low, which can be disadvantageous for dissolving the polar compounds found in biomass. However, CO_2 is very well suited for dissolving apolar and lipophilic molecules such as fats, oils and lipids considered as high valuable compounds in biomass processing. Although CO_2 has a limited miscibility with water, their combination leads to the *in situ* formation of carbonic acid inducting pH values around 3 at saturation, which is achieved by the double dissociation of water with CO_2 to form carbonic acid per the following equations:

$$CO_2 + 2H_2O \leftrightarrow HCO_3^- + H_3O^+, \; HCO_3^- + H_2O \leftrightarrow CO_3^{2-} + H_3O^+$$

A very interesting feature of this system is that pH values can be tuned by temperature and pressure allowing acid-catalyzed reaction rate enhancements and neutralization by depressurization. The pH values of CO_2–H_2O binary system can be estimated using the equation $pH = (8.00\times10^{-6})T^2 + 0.00209T - 0.216\ \ln(p_{CO_2}) + 3.92$, developed by van Walsum *et al.*, in the range of 100–250 °C and up to 150 atm in CO_2 partial pressure (T and P values should be in Celsius and atmospheres, respectively), which is calculated using Henry's Law.[34]

In addition, using a parameter known as reaction severity can facilitate the comparison of process conditions for different acid catalysts and temperatures by calculating the combined severity parameter (CS)[35] shown in equation $CS = \log(R_0) - \text{pH}$. These parameters can be particularly useful for comparing the addition/use of CO_2 during biomass processing to that of other acids and reaction mixtures. The combined severity parameter takes into consideration the acid concentration (pH values given above when CO_2 is used), the reaction time and the reaction temperature, which are used to calculate the severity factor, R_0. The severity factor is defined in equation $R_0 = t\exp\left(\frac{T-100}{14.75}\right)$ and was proposed by Overend *et al.*[36] for biomass hydrolysis in batch and plug-flow reactors (t and T values should be given in minutes and Celsius).

Given the attractive properties of this mixture, the use of high-pressure CO_2 and CO_2–H_2O systems has led to several improvements and developments in biomass processing.[5,37] In particular, CO_2 and CO_2–H_2O systems have led to promising results in areas involving high value compound extraction/purification, hydrolysis and dehydration of carbohydrates and biomass-derived hydrogenation. Table 2.1 lists several processes in which CO_2 has been used within biorefinery processes. Below, we highlight the principal applications and benefits of CO_2 and CO_2–H_2O mixtures in different types of biomass conversion technologies including physical and chemical processes.

2.4.2 Physical Processes Employing High-pressure CO_2 or CO_2–H_2O Systems

The utilization of CO_2 in physical processes has been extensively explored with a tremendous number of associated publications being available on the subject. Because of CO_2's ability to dissolve lipophilic substances,[38] the extraction of bioactive or high-value compounds using supercritical CO_2 ($scCO_2$) is one of the most common approaches used for the valorization and fractionation of biomass.[9,39–45] This approach is notably used extensively in industry especially for the production of food additives or medicinal products due to CO_2's non-toxic nature.[46] Within bioenergy applications, the targeted compounds during extractions have mainly been lipids, waxes, proteins, lignin and phenolic compounds for the production of biodiesel, fatty acids, amino acids, power/heat, aromatic compounds and other fine chemicals. The advantages of using CO_2 as an extraction or purification mediator over conventional processes such as organic solvent extraction and distillation include: reduction of operational steps, elimination of solvent waste, moderate operational temperatures, high quality of the extracts and cost saving while reducing negative impacts on the environment and human health risks.[38,47–49] $ScCO_2$ extraction has notably been extensively explored for the extraction of lipids from algae.[49–51] Several pilot scale studies and energy evaluation have been conducted to assess the performance of CO_2-based extractions.[50,51] Compared to organic solvent extractions, $scCO_2$ has shown higher efficiency and selectivity towards triglycerides avoiding the solubility of undesired components such as pigments or polar lipids.[49] The purification of biodiesel is an excellent example of the simplicity of CO_2-based processes; where, in a single injection step, biodiesel as pure as what is obtained by conventional purification was obtained but with much lower waste generation.[52] A similar use of CO_2 was described by Bourne *et al.* for the simultaneous production and purification of γ-valerolactone (GVL) from levulinic acid and water, where GVL was delivered almost pure with lower energy requirements than using a conventional reactor.[53] GVL is a biomass-derived precursor to several high-value chemicals and has excellent properties when used as a solvent for lignocellulosic biomass deconstruction and further upgrading reactions.[54–58] CO_2 can play a significant role in the recovery of GVL after reaction. Notably, liquid CO_2 was used to extract over 99% of GVL from the reaction media of several biomass conversion processes.[59,60] After its addition, a CO_2-expanded phase was created with GVL, which was no longer miscible with water. This lead to the production of a GVL phase and a concentrated aqueous phase containing over 90% of the carbohydrates produced from biomass which allowed for an easy recovery of GVL.[59–62]

Another important approach developed for biorefinery processes has been the use of CO_2 explosion to disrupt raw cellulosic substrates.[63–67] The instantaneous release of CO_2 at high pressure promotes the disruption of the cellulosic structure and leads to increases in the accessible surface area of the substrate used for further hydrolysis.[68] Processes that serve to increase accessible surface area in biomass play an especially key role for enzymatic hydrolysis. CO_2 has a similar ability to penetrate to accessible pores in biomass to water and ammonia ,which have comparable molecular sizes and are used in two well-known explosion treatments for biomass pre-treatment prior to enzymatic hydrolysis.[69,70] Some studies have reported that CO_2 explosion can be more cost-effective than ammonia explosion,[63] and can reduce the formation of inhibitor compounds produced during steam explosion through the degradation of sugars due to its lower operational temperatures.[68]

2.4.3 Chemical Processes Employing High-pressure CO_2 or CO_2–H_2O Systems

2.4.3.1 Thermochemical Processes

The addition of CO_2 in thermochemical processes has led to several beneficial effects on conversion, reaction rates, energy consumption and even selectivity. Butterman and Castaldi[71] reported that CO_2 enabled a more complete biomass gasification to volatiles. Notably, they reached an H_2/CO ratio suitable for FT fuel synthesis at a lower temperature than when using an N_2 environment. Also, a more efficient separation of lignin from holocellulose was possible. In the case of WT, the integration of CO_2 not only improved the conversion of biomass to volatiles but the resulting solids also showed an increase in heating value of up to 0.54 MJ kg^{-1} and a reduction of

Table 2.1 Examples of the principal CO_2-assisted processes proposed for use in biorefineries.

Main reactant	Process	Reaction media	Effect of CO_2 or CO_2–H_2O use	Source
Biomass [raw material]	Gasification	CO_2, $CO_2 + H_2O$	Higher yields to volatiles with a H_2/CO ratio more suitable for Fisher–Tropsch fuel synthesis, more efficient separation of lignin from holocellulose and the production of a very reactive and porous char at less severe conditions than in a N_2 environment.	71
	Wet torrefaction	$CO_2 + H_2O$	Higher ash removal, higher heating value, less specific gridding energy, reaction rate enhancement and less solid production due to the acidic catalysis effect compared to N_2-WT.	72
		$CO_2 + H_2O$	Higher ash removal, same hydrochar yield with a reduced heating value and less specific grinding energy compared to hydrothermal conditions.	73
	Pyrolysis	CO_2	Increase in CO production and six-fold increase in char surface area with a different chemical composition than chars produced under a N_2 environment. Inhibition of the secondary char formation and tar polymerization.	74
	Extraction	CO_2, $CO_2 + H_2O$; $CO_2 + EtOH$; CO_2 + organic alcohol or ether	Complete separation of cellulose and hemicellulose from lignin; fractionation of alkanes and fatty alcohols (waxes), extraction of phenolic compounds and α-mangostin; extraction of β-glucan from barley grains, precipitation of proteins from soy meal extracts or milk; extraction of a mixture of fatty acids, phenolic compounds, and fucoxanthin from *S. muticum*; extraction of lipids from living cells (microalgae); over 90% lignin removal from rice husk.	9, 39–45

	CO_2 explosion	CO_2, $CO_2 + H_2O$	Changes in chemical composition and crystallinity of the material with higher surface area and enhancement of enzymatic digestibility compared to the pure water treatment. Prevention of inhibitor produced in steam explosion and less sugar degradation due to lower temperature.	63–67
	Extraction and enzymatic transesterification	CO_2 + methanol–*tert*-butanol	Integrated extraction of fat from microalgae with no need for a solvent separation unit and the production of a high quality glycerol-free biodiesel.	48
	Acid hydrolysis	$CO_2 + H_2O$, high solid loading	Production of oligosaccharides and furfural from the hemicellulose fraction of biomass with a reduction in the crystallinity of the cellulose-rich fraction and higher susceptibility to enzymatic hydrolysis compared to hydrothermal conditions. Production of a highly concentrated monosaccharide solution after enzymatic hydrolysis. 33% less amount of enzyme required in later saccharification when CO_2 was used compared to diluted sulfuric acid pre-treatment.	76–81, 86
Cellulose	CO_2 explosion	CO_2 + buffer solution	Increase in accessible surface area of the cellulosic substrate and changes in the crystallinity of the material. Reduced production of inhibitors produced in steam explosion and less sugar degradation due to the lower temperature that is used with CO_2.	63, 66, 67
	Acid hydrolysis	$CO_2 + H_2O$	Enhancement in the hydrolysis rate constant compared to subcritical water treatment.	75
	Enzymatic hydrolysis	$CO_2 + H_2O$	Kinetic hydrolysis constant enhancement and retention of cellulase activity.	106–108

Table 2.1 (*Continued*)

Main reactant	Process	Reaction media	Effect of CO_2 or CO_2–H_2O use	Source
Starch	Acid hydrolysis	$CO_2 + H_2O$	Quasi-proportional increase in glucose yield from starch with the addition of CO_2. Hydrolysis rate constant increase.	75, 82
	Acid hydrolysis and dehydration	$CO_2 + H_2O$	Enhancement in dehydration and hydrolysis rates during the production of 5-HMF from inulin due to the formation of carbonic acid.	89
Hemicellulose or pentose	Acid hydrolysis, dehydration and extraction	$CO_2 + H_2O$	Production of furfural and its simultaneous extraction without the addition of acid.	90
Lignin	Depolymerization	$CO_2 + H_2O$ + acetone + HCOOH	Production of phenolic oil composed of oligomers fragments and aromatic monomers from lignin with slightly higher yields compared to catalytic steam processing.	99
	Purification	$CO_2 + H_2O$	Selective purification of lignin oxidation products from aqueous solution with lower energy requirements.	103
Protein	Acid hydrolysis	$CO_2 + H_2O$	Production of amino acids from bovine serum albumin with increased rates for peptide bond hydrolysis.	88
Polyalcohol	Dehydration	$CO_2 + H_2O$	Increase in THF yield from the dehydration of 1,4-butanediol. Enhancement in the dehydration rate of triol compounds compared to high-temperature liquid water. Increase in acetol yields due to dehydration rate enhancement and acetol stability compared to high-temperature liquid water.	83, 109, 110

Fatty acids	Hydrogenation	CO_2	Increase in the concentration of hydrocarbon compounds with diesel-like properties from cattle fat during hydrogenation.	91
Triglycerides/ Lipids	Transesterification/ Enzymatic transesterification	CO_2 + methanol, CO_2 + enzymes	Reduction in mass transfer limitations and increased catalytic active site exposure led to improvements in reaction. Reduction in the reaction temperature and time at 92% conversion.	96, 97
Levunic acid	Hydrogenation/ Purification	$CO_2 + H_2O$	Production and simultaneous purification of GVL from levulinic acid with lower energy requirements.	53
5-HMF or furfural	Hydrogenation	$CO_2 + H_2O$	Full conversion of 5-HMF and furfural to 2,5-DMF and 2-MF, respectively, with 100% selectivity. Tunable products profile by tuning the water : CO_2 ratios.	95
Biodiesel	Purification	CO_2	Purification of biodiesel in a single step without any loss in quality compared to conventional processes.	52

6.5 kWh t^{-1} in specific gridding energy.[72] However, the same research group also reported that the CO_2-enriched WT produced similar hydrochar yields with less heating value compared to N_2-WT, but with a significant increase in ash removal.[73] Similar to the impacts of CO_2 in biomass gasification, CO_2-assisted pyrolysis showed an enhancement in the production of CO as a result of reactions between CO_2 and other gases, tar and char.[74] The resulting char presented a six-fold increase in surface area with a different chemical composition than chars produced with a N_2 environment. It also appeared that CO_2 inhibited secondary char formation and tar polymerization.

2.4.3.2 *Catalytic Processes*

Some of the most versatile routes to produce bio-fuels and bio-products involve the upgrading of the carbohydrate fractions found in biomass. Given the large variations in biomass composition and chemical structure, the conversion technology that is employed generally has to be tailored to the biomass characteristics and to the subsequent upgrading routes to avoid low yields, high energy expenditures and excessive waste production.[5] For this reason, the initial biomass processing, including biomass pre-treatment, is one the most challenging and expensive steps in the biorefinery and has a significant influence on downstream processing. The utilization of CO_2–H_2O mixtures in the pre-treatment and hydrolysis media have several advantages compared to conventional processes such as the dilute acid or organosolv processes.[75–82] Notably, the formation of carbonic acid can enhance the rate of acid-catalyzed reactions, which include hydrolysis and dehydration, without the drawback of neutralization and/or solvent recovery.[75,76,82–86] The use of CO_2–H_2O mixtures during biomass pre-treatment has also been shown to reduce the crystallinity of the cellulose-rich fraction allowing a reduction in enzyme loading of 33% during subsequent saccharification.[86] Relvas *et al.* modelled the kinetics of the hemicellulose, xylan and arabinan hydrolysis in the presence of CO_2 added during hydrothermal processes and demonstrated that the formation of carbonic acid led to an increase in the kinetic reaction constants for both intermediates and final products.[87] The kinetics of cellulose and starch hydrolysis in CO_2–H_2O mixture have been the subject of similar studies and have demonstrated that CO_2 addition led to significant increases in reaction rates compared to pure water (see Figure 2.4 and Figure 2.5).[75] Other examples in which the *in situ* formation of carbonic acid have accelerated reaction rates include the hydrolysis of proteins for the production of amino acids,[88] and the dehydration of polyalcohols and monosaccharides from starch and hemicellulose for the production of tetrahydrofuran (THF),[83] 5-hydroxymethylfuran (5-HMF)[89] and furfural,[90] respectively. Further reaction examples accompanied with a discussion of the effects and advantages of CO_2 use are given in Table 2.1.

The hydrogenation or hydrodeoxygenation of low value feedstocks, such as cattle fat (rich in free fatty acids), is another pathway being considered for

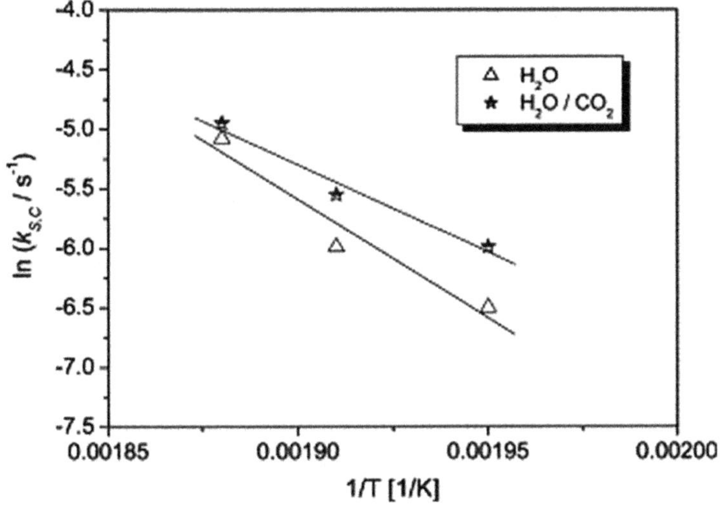

Figure 2.4 Arrhenius plots of cellulose hydrolysis in water and water–CO_2. Reprinted from T. Rogalinski, K. Liu, T. Albrecht and G. Brunner, *Hydrolysis Kinetics of Biopolymers in Subcritical Water*, vol. 46,[75] Copyright (2007) with permission from Elsevier.

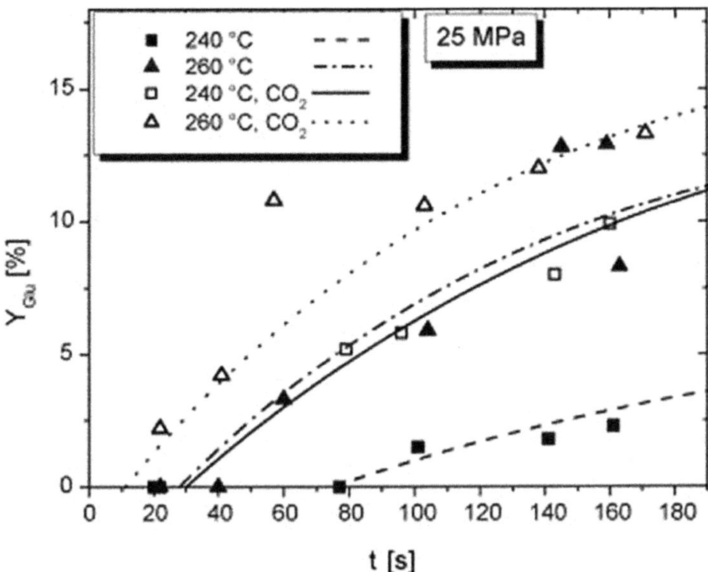

Figure 2.5 Glucose yields during subcritical water hydrolysis of cellulose at different residence times and temperatures in the presence and absence of dissolved CO_2.
Reprinted from T. Rogalinski, K. Liu, T. Albrecht and G. Brunner, *Hydrolysis Kinetics of Biopolymers in Subcritical Water*, vol. 46,[75] Copyright (2007) with permission from Elsevier.

biomass valorization.[91,92] However, several of these reactions can suffer from low reaction rates due to mass transfer limitations especially when they involve liquid-phase reactions. Notably, mass transfer limitations arise due to H_2 low solubility in common solvents. This leads to hydrogenation rates being controlled by the transfer of H_2 between the gas and liquid phase as opposed to the reaction itself. In contrast, CO_2 is highly miscible with most gases, which can greatly enhance gas solubility and any associated homogeneous or heterogeneous catalyzed reactions that suffer from gas-to-liquid mass transfer limitations.[93] In fact, the ability of CO_2 to enhance the solubility of several gases has been demonstrated in multiple reactions,[94] including several in biomass processing.[53,91,95] The rate of hydrodeoxygenation of cattle fat was enhanced by the addition of CO_2 which doubled the content of hydrocarbons in the final products, giving it diesel-like properties.[91] Interestingly, the presence of CO_2 induced a change in the catalyst's morphology leading to a reorganization of the kaolinite-based catalyst's internal structure in a similar way to what has been reported in literature when other processing effects such as flow, shear, electric polarization or colloids modified the catalyst properties.

Hydrogenation reactions also play important roles in several downstream upgrading routes proposed for biomass conversion including the production of several molecules proposed as fuel substitutes including 2,5-dimethyfuran (2,5-DMF) and 2-methylfuran (2-MF). 2,5-DMF and 2-MF are produced by the hydrogenation of 5-hydroxymethylfurfural (HMF) and furfural, which are derived from the dehydration of glucose and xylose, respectively. Recently, Chatterjee *et al.* reported that the addition of CO_2 into the reaction media led to full conversion of both substrates with 100% selectivity to the targeted products.[95] Figure 2.6 demonstrates how modifying the water/CO_2 ratio allows to tune the product distribution.

Also of importance for liquid biofuel production, the transesterification of triglycerides and methanol to fatty acid methyl esters (FAMEs) suffers from similar obstacles related to kinetics, selectivity and yields.[96] CO_2 was also used to enhance mass transfer between the substrates and catalyst leading to higher yields and modified product profiles. The reaction rates were also increased by the addition of CO_2 due to higher mass transfer but also because of CO_2's ability to swell the catalyst (the acid exchange resin, Nafion®) allowing greater exposure of the reactant to the active sites.[96] Transesterification reactions can also use enzymes as catalysts including lipases. The use of enzymes avoids side reactions and reduces the energy requirements involved in chemical methods due to the enzyme's near 100% selectivity and its ability to function at ambient or near ambient temperatures.[97] Lipases have been used in the presence of high-pressure CO_2 without denaturation for transesterification, hydrolysis and other reactions.[98] Hu *et al.* optimized the transesterification of phytosterol and soybean oil in $scCO_2$ using enzymes and achieved a conversion of 92% with significantly lower reaction temperatures and time than without CO_2.[97] The ability to use enzyme in CO_2–water mixtures has led to the integrated and continuous processing of

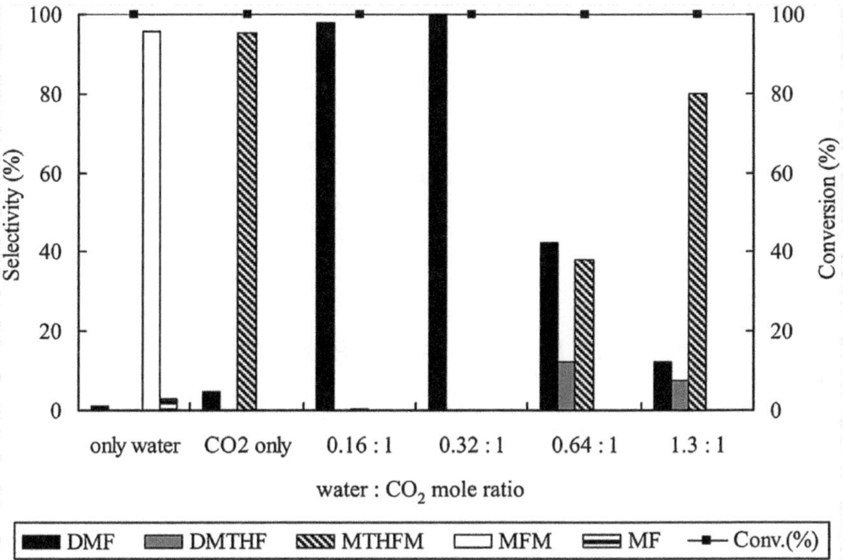

Figure 2.6 Effect of the variation of the water–CO_2 mole ratio on product distributions of furan hydrogenation. DMF = 2,5-dimethylfuran, DMTHF = 2,5-dimethyl-tetrahydro-furan; MTHFM = tetrahydro-5-methyl-2-furanmethanol, MF = 2,5-bis(hydroxymethyl)furan (MF), MFM = 5-methylfuranyl methanol.
Reproduced from ref. 95 with permission from The Royal Society of Chemistry.

microalgae lipids, including the extraction and biodiesel production steps *via* the enzymatic transesterification, in $scCO_2$.[48] The resulting biodiesel, dissolved in $scCO_2$, was easy to recover and its properties complied with the international standards and regulation without the need for further purification or glycerol separation.

The depolymerization and upgrading of lignin is also described as a key step in the production of liquid transportation fuel or aromatic molecules, and for furthering the economic development of lignocellulosic biorefineries.[99,100] However, the conversion of lignin to high-value chemicals and fuels presents many challenges due to its complex aromatic structure and poor chemical stability during depolymerization. Currently, 98% of lignin is burned for the production of heat and power in the pulp and paper industry.[101] Gosselink *et al.* studied the depolymerization of lignin to a phenolic oil (up to 45% based on lignin) in a supercritical fluid consisting of CO_2/acetone/water/formic acid.[99] The resulting phenolic oil was composed of oligomers fragments and contained several aromatic monomers corresponding to a 10–12% yield based on the original lignin. These monomers yields were slightly higher than those previously observed using catalytic steam processing.[102] Moreover, Assmann *et al.* showed that the selective extraction of lignin oxidative products is possible by changing the CO_2 pressure.[103] Through calculations, they suggested that the purification of lignin oxidative products using multi-stage CO_2 extractions could be used to produce single monomers at high purities while using a much more eco-friendly and cheap process compared to vacuum distillation and crystallization. In general, supercritical fluid extractions (SFE) require considerably less energy than conventional purification methods, but the operation costs of SFE is expensive.[104] Economic considerations in the case of vanillin purification have shown that, in order to be commercially competitive, the feed solution must contain at least 50% vanillin. This example illustrates the typical challenges of implementing SFE.

2.4.4 Challenges for Implementing Processes Using CO_2

In summary, high-pressure CO_2 and CO_2–H_2O systems provide the opportunity to engineer reaction conditions, reduce energy requirements and tune reaction rates, product selectivity and catalyst activity by manipulating only pressure, temperature or CO_2 content.[93] However, CO_2's low solvation of polar compounds requires the utilization of large quantities of CO_2[31] that can seriously compromises its industrial implementation. In fact, no integrated high-pressure CO_2 or CO_2–H_2O technologies currently exist in industrial biorefineries. Implementation concerns are related to high capital costs, the lack of specialized process engineers, low familiarity in CO_2-based processes advantages and safety issues in large scale operations.[52] These processes operate under high pressure and designing, building and controlling a high-pressure process needs specialized equipment and personnel.[31] However, though capital costs are often higher, the overall costs of these high-pressure processes are not always greater than their low-pressure alternatives. Daza Serna *et al.* compared the cost efficiency of the conventional acid pre-treatment with a $scCO_2$ pre-treatment for the conversion of rice husk and found that the $scCO_2$-based pre-treatment can be more cost effective leading to a reduction in the sugar production cost to 0.20 USD per kg from 1.88 USD per kg with a lower environmental impact.[9] A major fraction of the cost in high-pressure CO_2 processes is due to the high energy requirements for CO_2 conditioning (cooler and pump systems). However, the easy recovery of CO_2 and resulting high product concentrations represent opportunities for reducing the cost of raw material and energy savings during purification. Recently, Sharifzadeh *et al.* explored the economic feasibility and environmental performance of an integrated process utilizing CO_2 from biomass pyrolysis for the production biodiesel *via* microalgae.[105] The results showed a significant reduction of 45% to 6% of total carbon inputs in the flue gas and a potential increase of biomass to fuel yields from 55% to 73%, which compensated for the cost of CO_2 conditioning.

2.5 Conclusion

In this introduction, we discussed the general goals and possible routes for biomass conversion and biorefineries with attention to the possibilities of using high-pressure CO_2 and CO_2–H_2O. These high-pressure systems have shown to be beneficial in multiple steps during the biomass conversion chain and offer the added benefit of being non-toxic, green and renewable solvents. Furthermore, their use has been shown to benefit the production of first, second and third generation biofuels and bioproducts. Finally, when CO_2 is used and recycled in confined processes it can be considered as "sequestered" if its source is natural deposits or man-made production. Some of the principal benefits associated with CO_2 use include:

i. Acid-catalyzed reaction rate enhancements
ii. Tunability of the product distribution
iii. High products concentration after CO_2
iv. Increase in catalysts activity through solvent effects or catalyst swelling
v. Reduction of mass transfer limitations
vi. Milder process conditions
vii. Simplified efficient purification/extraction steps
viii. Lower environmental impact through reduced waste production.

Although the cost-effectiveness of high-pressure technologies over conventional processes is not systematic and must be evaluated for each individual process, the advantages associated with the use of CO_2 can help compensate for the energy and capital requirements of high pressure processing and CO_2 use. Nevertheless, industrial implementation is still hindered by the lack of specialized process engineers and safety concerns over large-scale high pressure operations. Therefore, additional fundamental understanding of high-pressure CO_2 and CO_2–water systems combined with detailed techno-economic modelling is needed to ensure an optimal process design for each application and facilitate the beneficial use of CO_2 in future biorefineries.

References

1. J. Luterbacher, D. M. Alonso and J. Dumesic, *Green Chem.*, 2014, **16**, 4816–4838.
2. S. Sorrell, *Renewable Sustainable Energy Rev.*, 2015, **47**, 74–82.
3. Key World Energy Statistics, Paris, France, 2015.
4. A. Eisentraut, A. Brown and L. Fulton, *Technology Roadmap: Biofuels for Transport*, 2011.
5. A. R. C. Morais, A. M. da Costa Lopes and R. Bogel-Lukasik, *Chem. Rev.*, 2015, **115**, 3–27.
6. C. Anton, *Bioenergy-Chances and Limits: Statement*, German National Academy of Sciences, Leopoldina, 2012.
7. F. Cherubini, *Energy Convers. Manage.*, 2010, **51**, 1412–1421.
8. *Standard Terminology Relating to Biotechnology*, ASTM International, West Conshohocken, PA, 2015.

9. L. D. Serna, C. O. Alzate and C. C. Alzate, *Bioresour. Technol.*, 2016, **199**, 113–120.
10. *Renewable fuel standard: potential economic and environmental effects of US biofuel policy*, Report 0309187516, National Research Council. Committee on Economic Environmental Impacts of Increasing Biofuels Production 2011.
11. J. C. Sinistore, D. J. Reinemann, R. C. Izaurralde, K. R. Cronin, P. J. Meier, T. M. Runge and X. Zhang, *BioEnerg. Res.*, 2015, **8**, 897–909.
12. D. M. Alonso, J. Q. Bond and J. A. Dumesic, *Green Chem.*, 2010, **12**, 1493–1513.
13. A. Ahmad, N. M. Yasin, C. Derek and J. Lim, *Renewable Sustainable. Energy Rev.*, 2011, **15**, 584–593.
14. B. Subramaniam, R. V. Chaudhari, A. S. Chaudhari, G. R. Akien and Z. Xie, *Chem. Eng. Sci.*, 2014, **115**, 3–18.
15. S. Ahmed, A. Jaber and R. Dixon, *Renewables 2010 Global Status Report: Renewable Energy Policy Network for the 21st Century*, REN.
16. G. P. Hammond and B. Li, *Environmental and resource burdens associated with world biofuel production out to 2050: footprint components from carbon emissions and land use to waste arisings and water consumption*, Report 1757-1707, 2015.
17. A. Wellinger, J. D. Murphy and D. Baxter, *The Biogas Handbook: Science, Production and Applications*, Elsevier, 2013.
18. J. A. Onwudili, in *Application of Hydrothermal Reactions to Biomass Conversion*, ed. F. Jin, Springer, 2014, pp. 219–246.
19. Y. Chisti, *Biotechnol. Adv.*, 2007, **25**, 294–306.
20. T. M. Mata, A. A. Martins and N. S. Caetano, *Renewable Sustainabale Energy Rev.*, 2010, **14**, 217–232.
21. F. Kaup, *The Sugarcane Complex in Brazil: The Role of Innovation in a Dynamic Sector on Its Path Towards Sustainability*, Springer, 2015.
22. J. v. Haveren, E. L. Scott and J. Sanders, *Biofuels, Bioprod. Biorefin.*, 2008, **2**, 41–57.
23. T. Werpy and G. R. Petersen, *Top Value Added Chemicals From Biomass - Volume I: Results of Screening for Potential Candidates from Sugars and Synthesis Gas* the Pacific Northwest National Laboratory (PNNL) and the National Renewable Energy Laboratory (NREL), 2004.
24. *Thermochemical value chains for production of biofuels*, European Biofuels Technology Platform, Brussels, Belgium, 2016.
25. J. S. Luterbacher, M. Fröling, F. Vogel, F. Maréchal and J. W. Tester, *Environ. Sci. Technol.*, 2009, **43**, 1578–1583.
26. M. E. Dry, *Catal. Today*, 2002, **71**, 227–241.
27. N. Tien-Thao, M. H. Zahedi-Niaki, H. Alamdari and S. Kaliaguine, *Appl. Catal., A*, 2007, **326**, 152–163.
28. J. Sun, S. Wan, F. Wang, J. Lin and Y. Wang, *Ind. Eng. Chem. Res.*, 2015, **54**, 7841–7851.
29. *Torrefaction of biomass and "biocoal" technologies*, European Biofuels Technology Platform, Brussels, Belgium, 2016.

30. Q.-V. Bach, K.-Q. Tran, R. A. Khalil, Ø. Skreiberg and G. Seisenbaeva, *Energ. Fuel.*, 2013, **27**, 6743–6753.
31. Y. Medina-Gonzalez, S. Camy and J.-S. Condoret, *Int. J. Sustain. Eng.*, 2012, **5**, 47–65.
32. H. G. Niessen and K. Woelk, Investigations in Supercritical Fluids, in *In situ NMR Methods in Catalysis*, ed. J. Bargon and L. T. Kuhn, Springer Berlin Heidelberg, Germany, 2005, vol. **276**, ch. 3, pp. 69–110.
33. E. J. Beckman, *J. Supercrit. Fluid.*, 2004, **28**, 121–191.
34. G. P. van Walsum, *Appl. Biochem. Biotechnol.*, 2001, **91–93**, 317–329.
35. H. L. Chum, D. K. Johnson, S. K. Black and R. P. Overend, *Appl. Biochem. Biotechnol.*, 1990, **24–5**, 1–14.
36. R. P. Overend, E. Chornet and J. A. Gascoigne, *Philos. Trans. R. Soc., A*, 1987, **321**, 523–536.
37. R. A. Sheldon, *Green Chem.*, 2014, **16**, 950–963.
38. F. Sahena, I. Zaidul, S. Jinap, A. Karim, K. Abbas, N. Norulaini and A. Omar, *J. Food Eng.*, 2009, **95**, 240–253.
39. F. E. I. Deswarte, J. H. Clark, J. J. E. Hardy and P. M. Rose, *Green Chem.*, 2006, **8**, 39–42.
40. K. Chhouk, A. T. Quitain, D. G. Pag-asa, J. B. Maridable, M. Sasaki, Y. Shimoyama and M. Goto, *J. Supercrit. Fluid.*, 2016, **110**, 167–175.
41. S. Kodama, T. Shoda, S. Machmudah, H. Kanda and M. Goto, *Chem. Eng. Process.*, 2015, **97**, 45–54.
42. N. Khorshid, M. M. Hossain and M. Farid, *J. Food Eng.*, 2007, **79**, 1214–1220.
43. E. Conde, A. Moure and H. Domínguez, *J. Appl. Phycol.*, 2015, **27**, 957–964.
44. A. Zinnai, C. Sanmartin, I. Taglieri, G. Andrich and F. Venturi, *J. Supercrit. Fluid.*, 2016, **116**, 126–131.
45. F. Maoqi, D. D. Daruwalla, J. Erwin and W. K. Gauger, US9217119, 2015.
46. N. T. Dunford, *Food and Industrial Bioproducts and Bioprocessing*, John Wiley & Sons, 2012.
47. M. Djas and M. Henczkab, *J. Supercrit. Fluid.*, 2016, **117**, 59–63.
48. H. Taher, S. Al-Zuhair, A. H. Al-Marzouqi, Y. Haik and M. Farid, *World Acad. Sci. Eng. Technol. Int. J. Environ. Chem. Ecol. Geol. Geophys. Eng.*, 2015, **9**, 884–887.
49. L. Soh and J. Zimmerman, *Green Chem.*, 2011, **13**, 1422–1429.
50. W. Bjornsson, K. MacDougall, J. Melanson, S. O'Leary and P. McGinn, *J. Appl. Phycol.*, 2012, **24**, 547–555.
51. Y. Du, B. Schuur, S. R. Kersten and D. W. Brilman, *Algal Res.*, 2015, **11**, 271–283.
52. A. M. Escorsim, C. S. Cordeiro, L. P. Ramos, P. M. Ndiaye, L. R. Kanda and M. L. Corazza, *J. Supercrit. Fluid.*, 2015, **96**, 68–76.
53. R. A. Bourne, J. G. Stevens, J. Ke and M. Poliakoff, *Chem. Commun.*, 2007, 4632–4634.
54. M. A. Mellmer, C. Sener, J. M. R. Gallo, J. S. Luterbacher, D. M. Alonso and J. A. Dumesic, *Angew. Chem., Int. Ed.*, 2014, **53**, 11872–11875.

55. M. A. Mellmer, D. M. Alonso, J. S. Luterbacher, J. M. R. Gallo and J. A. Dumesic, *Green Chem.*, 2014, **16**, 4659–4662.
56. D. M. Alonso, S. G. Wettstein, M. A. Mellmer, E. I. Gurbuz and J. A. Dumesic, *Energy Environ. Sci.*, 2013, **6**, 76–80.
57. D. M. Alonso, S. G. Wettstein and J. A. Dumesic, *Green Chem.*, 2013, **15**, 584–595.
58. D. M. Alonso, J. M. R. Gallo, M. A. Mellmer, S. G. Wettstein and J. A. Dumesic, *Catal. Sci. Technol.*, 2013, **3**, 927–931.
59. J. Han, J. S. Luterbacher, D. M. Alonso, J. A. Dumesic and C. T. Maravelias, *Bioresour. Technol.*, 2015, **182**, 258–266.
60. L. Shuai, Y. M. Questell-Santiago and J. S. Luterbacher, *Green Chem.*, 2016, **18**, 937–943.
61. J. S. Luterbacher, D. M. Alonso, J. M. Rand, Y. M. Questell-Santiago, J. H. Yeap, B. F. Pfleger and J. A. Dumesic, *ChemSusChem*, 2015, **8**, 1317–1322.
62. J. S. Luterbacher, J. M. Rand, D. M. Alonso, J. Han, J. T. Youngquist, C. T. Maravelias, B. F. Pfleger and J. A. Dumesic, *Science*, 2014, **343**, 277–280.
63. Y. Zheng, H. Lin and G. T. Tsao, *Biotechnol. Prog.*, 1998, **14**, 890–896.
64. N. Narayanaswamy, A. Faik, D. J. Goetz and T. Y. Gu, *Bioresour. Technol.*, 2011, **102**, 6995–7000.
65. H. D. Zhang and S. B. Wu, *Bioresour. Technol.*, 2014, **158**, 161–165.
66. Y. Zheng, H.-M. Lin, J. Wen, N. Cao, X. Yu and G. T. Tsao, *Biotechnol. Lett.*, 1995, **17**, 845–850.
67. A. L. F. Santos, K. Y. F. Kawase and G. L. V. Coelho, *J. Supercrit. Fluid.*, 2011, **56**, 277–282.
68. P. Kumar, D. M. Barrett, M. J. Delwiche and P. Stroeve, *Ind. Eng. Chem. Res.*, 2009, **48**, 3713–3729.
69. C. Sarks, B. D. Bals, J. Wynn, F. Teymouri, S. Schwegmann, K. Sanders, M. Jin, V. Balan and B. E. Dale, *Biofuels*, 2016, 1–15.
70. S. Duangwang, T. Ruengpeerakul, B. Cheirsilp, R. Yamsaengsung and C. Sangwichien, *Bioresour. Technol.*, 2016, **203**, 252–258.
71. H. C. Butterman and M. J. Castaldi, *Environ. Sci. Technol.*, 2009, **43**, 9030–9037.
72. Q.-V. Bach, K.-Q. Tran, Ø. Skreiberg and R. A. Khalil, *Energy Procedia*, 2014, **61**, 1200–1203.
73. Q.-V. Bach, K.-Q. Tran and Ø. Skreiberg, *Fuel Process. Technol.*, 2015, **140**, 297–303.
74. C. Guizani, F. E. Sanz and S. Salvador, *Fuel*, 2014, **116**, 310–320.
75. T. Rogalinski, K. Liu, T. Albrecht and G. Brunner, *J. Supercrit. Fluid.*, 2008, **46**, 335–341.
76. S. P. Magalhães da Silva, A. R. C. Morais and R. Bogel-Lukasik, *Green Chem.*, 2014, **16**, 238–246.
77. J. S. Luterbacher, J. W. Tester and L. P. Walker, *Biotechnol. Bioeng.*, 2012, **109**, 1499–1507.

78. F. M. Relvas, A. R. C. Morais and R. Bogel-Lukasik, *RSC Adv.*, 2015, 73935–73944.
79. J. S. Luterbacher, J. W. Tester and L. P. Walker, *Biotechnol. Bioeng.*, 2010, **107**, 451–460.
80. A. R. C. Morais, A. C. Mata and R. Bogel-Lukasik, *Green Chem.*, 2014, **16**, 4312–4322.
81. J. S. Luterbacher, Q. Chew, Y. Li, J. W. Tester and L. P. Walker, *Energy Environ. Sci.*, 2012, **5**, 6990–7000.
82. T. Miyazawa and T. Funazukuri, *Biotechnol. Prog.*, 2005, **21**, 1782–1785.
83. O. Sato, A. Yamaguchi and M. Shirai, *Catal. Commun.*, 2015, **68**, 6–10.
84. S. E. Hunter and P. E. Savage, *Ind. Eng. Chem. Res.*, 2003, **42**, 290–294.
85. G. P. van Walsum and H. Shi, *Bioresour. Technol.*, 2004, **93**, 217–226.
86. J. W. King, K. Srinivas, O. Guevara, Y. W. Lu, D. F. Zhang and Y. J. Wang, *J. Supercrit. Fluid.*, 2012, **66**, 221–231.
87. F. M. Relvas, A. R. C. Morais and R. Bogel-Lukasik, *J. Supercrit. Fluid.*, 2015, **99**, 95–102.
88. T. Rogalinski, S. Herrmann and G. Brunner, *J. Supercrit. Fluid.*, 2005, **36**, 49–58.
89. S. X. Wu, H. L. Fan, Y. Xie, Y. Cheng, Q. A. Wang, Z. F. Zhang and B. X. Han, *Green Chem.*, 2010, **12**, 1215–1219.
90. K. Gairola and I. Smirnova, *Bioresour. Technol.*, 2012, **123**, 592–598.
91. A. R. C. Morais, A. M. da Costa Lopes, P. Costa, I. Fonseca, I. N. Nogueira, A. C. Oliveira and R. Bogel-Lukasik, *RSC Adv.*, 2014, **4**, 32081–32091.
92. A. R. C. Morais, A. M. D. Lopes and R. Bogel-Lukasik, *Monatsh. Chem.*, 2014, **145**, 1555–1560.
93. C. M. Wai, F. Hunt, M. Ji and X. Chen, *J. Chem. Educ.*, 1998, **75**, 1641.
94. M. G. Hitzler and M. Poliakoff, *Chem. Commun.*, 1997, 1667–1668.
95. M. Chatterjee, T. Ishizaka and H. Kawanami, *Green Chem.*, 2014, **16**, 1543–1551.
96. L. Soh, C.-C. Chen, T. A. Kwan and J. B. Zimmerman, *ACS Sustainable Chem. Eng.*, 2015, **3**, 2669–2677.
97. L. Hu, S. Llibin, J. Li, L. Qi, X. Zhang, D. Yu, E. Walid and L. Jiang, *Bioprocess Biosyst. Eng.*, 2015, **38**, 2343–2347.
98. M. Jackson, J. King, G. List and W. Neff, *J. Am. Oil Chem. Soc*, 1997, **74**, 635–639.
99. R. J. A. Gosselink, W. Teunissen, J. E. G. van Dam, E. de Jong, G. Gellerstedt, E. L. Scott and J. P. M. Sanders, *Bioresour. Technol.*, 2012, **106**, 173–177.
100. J. E. Holladay, J. J. Bozell, J. F. White and D. Johnson, DOE Report PNNL-16983 (*website*: http://chembioprocess.pnl.gov/staff/staff_info.asp), 2007.
101. R. Behling, S. Valange and G. Chatel, *Green Chem.*, 2016, **18**, 1839–1854.
102. J.-M. Lavoie, W. Baré and M. Bilodeau, *Bioresour. Technol.*, 2011, **102**, 4917–4920.

103. N. Assmann, H. Werhan, A. Ładosz and P. R. von Rohr, *Chem. Eng. Sci.*, 2013, **99**, 177–183.
104. M. I. F. Mota, P. C. Rodrigues Pinto, J. M. Loureiro and A. E. Rodrigues, *Sep. Purif. Rev.*, 2016, **45**, 227–259.
105. M. Sharifzadeh, L. Wang and N. Shah, *Renewable Sustainable Energy Rev.*, 2015, **47**, 151–161.
106. G. Muratov, *Chem. Nat. Compd.*, 2007, **43**, 641–642.
107. C. Y. Park, Y. W. Ryu and C. Kim, *Korean J. Chem. Eng.*, 2001, **18**, 475–478.
108. Y. Z. Zheng and G. T. Tsao, *Biotechnol. Lett.*, 1996, **18**, 451–454.
109. A. Yamaguchi, N. Hiyoshi, O. Sato, C. V. Rode and M. Shirai, *Chem. Lett.*, 2008, **37**, 926–927.
110. A. Yamaguchi, N. Hiyoshi, O. Sato and M. Shirai, *Top. Catal.*, 2010, **53**, 487–491.

CHAPTER 3

Pre-treatment of Biomass Using CO_2-based Methods

LUIZ P. RAMOS,*[a] FAYER M. DE LEÓN MAYORGA,[a] MARCOS H. L. SILVEIRA,[b] CÉLIA M. A. GALVÃO[b] AND MARCOS L. CORAZZA[c]

[a] CEPESQ, Centro de Pesquisa em Química Aplicada, Department of Chemistry, Federal University of Paraná, P.O. Box 19032, 81531-980 Curitiba, Paraná, Brazil; [b] CTC, Cane Technology Center, P.O. Box 162, 13400-970 Piracicaba, São Paulo, Brazil; [c] Department of Chemical Engineering, Federal University of Paraná, 81531-990 Curitiba, Paraná, Brazil

*Email: luiz.ramos@ufpr.br

3.1 Introduction

Lignocellulosic materials are suitable feedstocks for the production of fuels, chemicals and materials, being a promising alternative to non-renewable precursors such as crude oil, coal and natural gas. However, for the successful conversion of lignocellulosics, a pre-treatment process must be developed to open up the structure of the plant cell wall and expose the plant polysaccharides to further chemical, biological or thermo-mechanical conversion.[1–3] This process is not simple, due to the close association that exists among the three main components of the plant cell wall, cellulose, hemicelluloses and lignin; also, the polyphenolic nature of lignin acts as a physical barrier to both chemical and biochemical transformations of plant polysaccharides. Therefore, pre-treatment must reduce the interactions

Green Chemistry Series No. 48
High Pressure Technologies in Biomass Conversion
Edited by Rafał M. Łukasik

Published by the Royal Society of Chemistry, www.rsc.org

among these macromolecular components, promote high recovery yields of biomass components, increase the chemical accessibility of plant polysaccharides and reduce the generation of inhibitory compounds for conversion processes based on enzymatic hydrolysis and fermentation.[4–6] In addition, a suitable pre-treatment method must have a low energetic demand, operate at high total solids, and be environmentally acceptable under the current criteria of process sustainability.

In general, pre-treatment represents one of the most crucial stages for the development of biorefineries based on biomass conversion processes. One such process, presently at its pre-commercialization stage, is the production of cellulosic ethanol for fuel applications. In this case, pre-treatment is responsible for increasing the susceptibility of plant polysaccharides to enzymatic hydrolysis (mostly glucans) but this must be done without promoting the accumulation of by-products such as organic acids, phenolics and furan compounds because these may be highly inhibitory to yeast fermentation to produce cellulosic ethanol.[3]

Different pre-treatment methods have been proposed to date for the fractionation of lignocellulosic materials in different process streams. Such methods are based on mechanical, physical, chemical or biological transformation of the plant cell wall structure but the most successful ones are those that combine at least two of these pre-treatment strategies. The most successful ones involve the use of acid hydrolysis,[7] steam explosion,[8] alkaline extraction,[9,10] wet oxidation,[11] heteropolyacids,[12] liquid ammonia (Ammonia Fibre Explosion or AFEX),[13] hot water extraction,[14] hydrothermal conversion,[15] milling,[16] ozonolysis,[17] ionic liquids,[18,19] organic solvents,[20–22] biopulping,[23] ultrasound,[24] microwave irradiation[25] and supercritical fluids,[26,27] all of these with the purpose of deconstructing the associative structure of the plant cell wall for the production of renewable fuels, chemicals and materials.

Steam explosion is one of the most efficient methods for the fractionation of lignocellulosic materials because it is based on both physical (including thermal) and chemical transformations of the plant cell wall. When the biomass is submitted to this type of pre-treatment, an increase in the accessible surface area is achieved with a consequent increase in the susceptibility of plant polysaccharides to enzymatic hydrolysis and this is mostly due to changes in hemicellulose and lignin structures that are triggered by acid hydrolysis.[8,28] This technology has a relatively low capital cost, low environmental impact, favourable energy balance and high sugar recoveries for the subsequent conversion steps.[6] In general, high-pressure steaming releases acetic acid by acid hydrolysis of acetylated hemicelluloses as well as levulinic and formic acids by rehydration of furan compounds such as 5-hydroxymethylfurfural (HMF), which are formed by dehydration of hexoses at higher pre-treatment severities.[8] These organic acids released *in situ* are able to catalyze the acid hydrolysis of acid-labile glycosidic linkages in a process called autohydrolysis.[29] However, depending of the time and temperature used for pre-treatment, these reactions continue to produce more dehydration by-products and cause deeper modifications in the structure of lignin that usually compromise its further use in chemical processes. Such modifications are mainly attributed to oxidation and condensation reactions involving phenolics and carbohydrate derivatives whose detrimental effects depend on the pre-treatment severity. Also, when the pressure of the reaction chamber is released (explosion or decompression), the fibre bundles are broken and split apart due to a strong shearing effect against the reactor inner walls.[30] Steam explosion can also be carried out in the presence of exogenous catalysts or impregnation agents such as dilute mineral acid,[31] SO_2,[32] CO_2,[33] NH_3,[34] Lewis acids[35] and dilute alkaline hydroxides such as NaOH.[10]

Acid catalysts are usually used to enhance the pre-treatment selectivity towards hemicellulose removal, which is recovered mostly as monomers. Also, acid catalysis induces greater modifications in the lignin structure, which may enhance the accessibility of glucans to chemical or biological (including enzymatic) conversions, and reduce the temperature and time required for optimal pre-treatment, possibly generating less dehydration by-products and/or inhibitory compounds.[31,36]

Mild-acid hydrolysis can be a selective and non-aggressive way to remove the hemicellulose component, making cellulose more accessible to its subsequent processing.[37] Also, the use of weak acids as pre-treatment catalysts have a less destructive effect on plant polysaccharides and promote less corrosion and less accumulation of furan compounds (furfural and HMF) in the reaction media.[38] However, the resulting sugar yields would depend on the biomass chemistry and particle size as well as on process parameters such as reaction temperature, time, and type and concentration of the acid catalyst.

Several weak acids have been proposed to assist the thermochemical pre-treatment of lignocellulosic materials. These include both mineral and organic acids such as phosphoric, acetic and oxalic acids, among others.[3] Compared to strong acids such as sulfuric acid, the use of dilute phosphoric acid as a pre-treatment catalyst results in lower sugar losses, less accumulation of furan compounds in the reaction medium, lower inhibitory effect on microbial growth and less corrosion of valves, pipelines and reactor inner walls.[37,39,40] However, phosphoric acid is more expensive than sulfuric acid and must not be discarded in waste waters because it may interfere with the phosphorus cycle and accelerate eutrophication.[37]

Organic acids have been used for biomass pre-treatment as well. For instance, aqueous mixtures of formic and acetic acids were used to pre-treat *Miscanthus x giganteus* biomass in different volumetric ratios.[41] The best pre-treatment condition was achieved at 107 °C for 3 h using a formic acid–acetic acid–water ratio of 40 : 40 : 20, which gave a glucan yield by enzymatic hydrolysis of 75.3%. In general, the pre-treatment temperature had the greatest effect on biomass delignification and this was closely related to the development of glucan susceptibility to hydrolysis. The use of 2–10 wt% oxalic acid, in temperatures ranging from 170 to 190 °C and reaction times of

15 to 40 min, removed most of the hemicellulose component of giant reed, leaving a solid residue that was highly accessible to enzymatic hydrolysis.[42]

Steam explosion has also been carried out in the presence of a mixture of one strong mineral acid and one weak organic acid. Sulfuric and acetic acids were used in two different solid-to-liquid ratios (1.5 : 10 and 1 : 10) to produce fermentable sugars from sugarcane bagasse. The hemicellulose content of native bagasse was reduced by 90%, the lignin component was only slightly affected and glucan losses were below 15%.[43] In general, these sugar yields were similar to those obtained with other traditional methods of acid hydrolysis.

In general, the acid strength is a key issue to define the selectivity of a pre-treatment process based on acid hydrolysis. Two other important criteria for the viability of a sustainable pre-treatment method are the achievement of a favourable energy balance and a low environmental impact, which include the need for producing highly accessible substrates for hydrolysis and low sugar losses to enzyme/fermentation inhibitors. Therefore, the tunability and wide applications of supercritical carbon dioxide ($scCO_2$) justify its proposition as a greener technology for the pre-treatment of different biomass types. The following section describes the role of CO_2 as a tool for enhancing the chemical and/or biological conversion of lignocellulosic materials into fuels, chemicals and materials.

3.2 CO_2 Properties Under Subcritical and Supercritical Conditions

A fluid is defined at supercritical condition when it is maintained at conditions above its critical temperature and pressure (T_c and P_c), where it exhibits properties of liquids such as density and gases such as compressibility. CO_2 is the most widely used supercritical fluid due to its non-toxic, recyclable, low-cost, environmentally friendly characteristics (when considered in a close loop process), and low critical temperature and pressure (31.1 °C and 7.38 MPa, respectively).

Supercritical and subcritical fluids have been used in pre-treatments since late eighties.[44,45] The most important parameters in biomass pre-treatment with sub- or supercritical fluids (SCFs) are the moisture, temperature, pressure and pre-treatment time. Water in the biomass combined with $scCO_2$ generates carbonic acid and this weak acidic environment prompts hemicellulose hydrolysis, while the SCF intensifies the mass transfer conditions.[26,46] Additionally, when water is mixed with other solvents the delignification capability is enhanced; for instance, ethanol improves the solubility of lignin fragments.[47]

According to the literature,[26,48,49] the high-pressure condition enhances the interaction between fluids or solvent mixtures and lignocellulosic biomass promoting the rupture of the bonds between lignin and the lignocellulosic matrix.

Even though the capability of SCFs to act as solvents has been known over a century, the most significant developments in the science and engineering applications of this technology have taken place in the last three decades. Among these, SCFs have been employed as media for chemical reactions and as solvent for extraction processes and, lately, as alternative for biomass pre-treatment. As a reaction medium, the SCF may either participate directly in the reaction or simply act as a solvent for the various chemical species. The physical properties of SCFs are highly dependent on pressure and temperature, and such dependence makes it possible to fine tune the properties of the extraction/reaction environment.[3,50] In view of their importance to process design, fundamental aspects of some thermodynamic properties are described in the following section, aiming to address the understanding of process application with pure $scCO_2$ and its binary mixtures with organic solvents (co-solvents).

3.2.1 Physicochemical Properties and Phase Behaviour of CO_2

The knowledge about the CO_2 phase behaviour and how its main thermodynamic and transport properties vary with temperature and pressure is a fundamental task in any application with SCFs. In Figure 3.1(A,B), a general pressure-density-temperature diagram of CO_2 is shown that was generated from the NIST web-book.[51] The critical point is given by the red open triangle and the vapour–liquid two phase region is given by the continuous black line (also referred as saturation line), while the dashed red line in Figure 3.1(B) represents the critical isotherm. From these diagrams the critical density ρ_c is defined as 467.7 kg m^{-3}, the critical temperature (T_c) is 30.9782 °C and the critical pressure (P_c) is 7.3773 MPa. In Figure 3.1 as well as in other figures included in this work, only the vapour and liquid states of CO_2 are shown to better define the supercritical region; the solid region was not included just for the sake of clarity.

Figure 3.2 presents different isotherms in a pressure–density–temperature diagram for CO_2, whose shape change considerably at temperatures above the critical temperature. Also, the CO_2 density changes sharply with small variations in one variable of the conjugated pair temperature–pressure, while keeping the other one fixed. This is one of the most important characteristic of using $scCO_2$ because, by tuning the density, the solvent power of $scCO_2$ is set, reaching different solubility parameter and features. It is important to notice that specific densities of $scCO_2$, such as $\rho_r = \rho/\rho_c = 1.2$, can be achieved at several different values of the pressure-temperature coordinates.

The properties of SCFs are very sensitive to small changes in temperature and pressure in the vicinity of the critical point. This applies especially to density as shown in Figure 3.2, which represents the projection of the liquid–vapour coexistence curve in the pressure–density space. The lower envelop of the diagram (saturated liquid and vapour lines) reveals the two phase

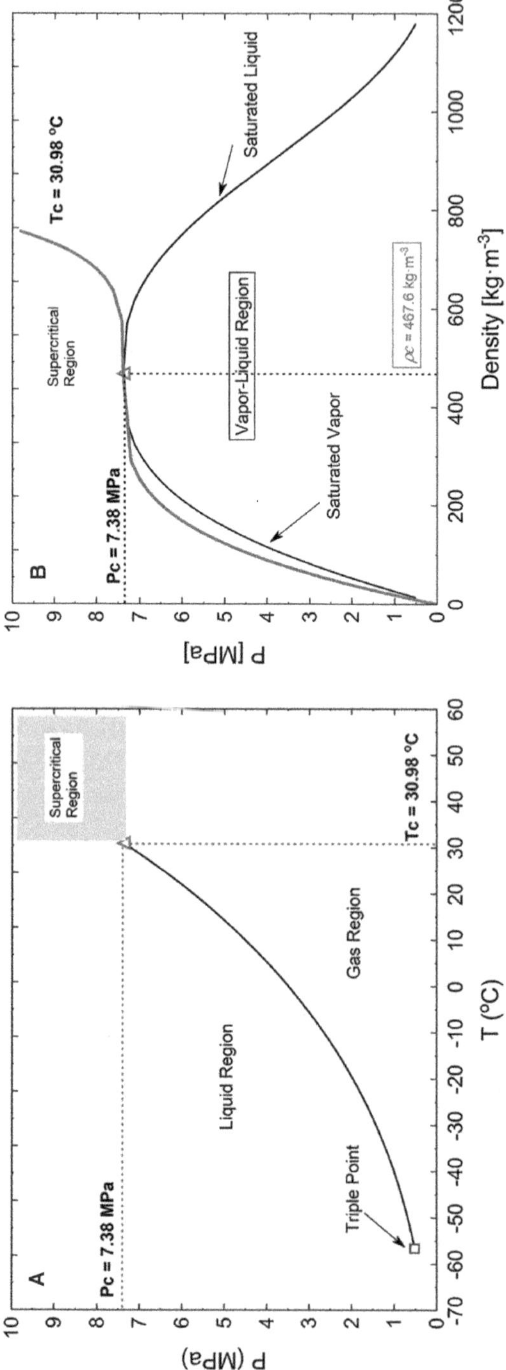

Figure 3.1 Pressure–temperature (A) and pressure–density (B) projections of CO_2.

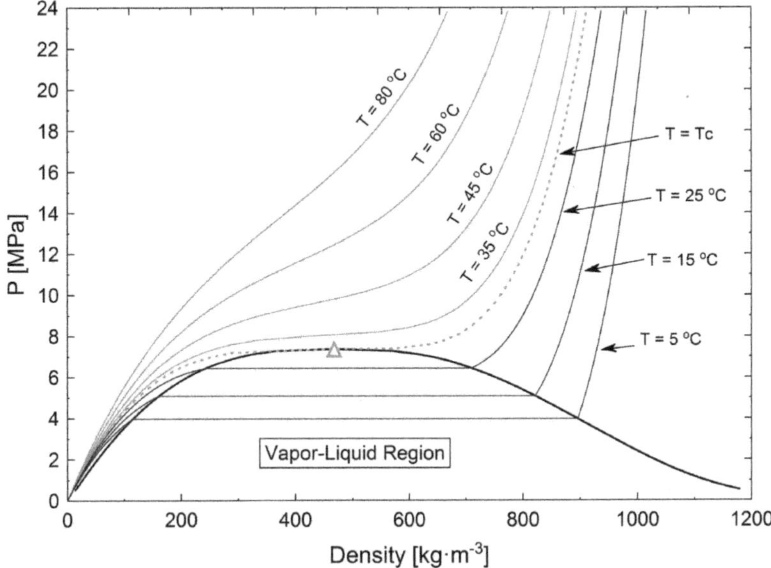

Figure 3.2 Pressure–density diagram of CO_2 at different isotherms. $T=35–80$ °C and $T=5–25$ °C are indications of temperatures below and above the critical temperature, respectively, and the dashed line is the critical isotherm. The triangle is the critical point.

envelope and the equilibrium between liquid and vapour phases. The supercritical region is defined by the region at isotherms of $T \geq T_c$, $P \geq P_c$ and $\rho \geq \rho_c$.

In the immediate vicinity of the critical point, the density of $scCO_2$ is around 400 kg m^{-3}. For reduced temperatures ($T\ T_c^{-1}$) greater than 1.0, the density of $scCO_2$ can be turned to values comparable to that of liquid CO_2 by simply increasing the pressure. It is this liquid-like density that improves the $scCO_2$ solvent strength to a level that is several orders of magnitude greater than what is predicted by ideal gas considerations. Since density is a measure of the solvating power of a SCF, temperature and pressure can be used as variables to control the solubility and the separation of a solute.

Following the same analysis, Figure 3.3 and Figure 3.4 present other important properties that must be considered when analyzing and/or working with $scCO_2$ applications such as biomass pre-treatment. Figure 3.3 presents a pressure–viscosity diagram of CO_2 and Figure 3.4 shows a pressure–speed of sound diagram at different isotherms.

The diffusion coefficient (or diffusivity) and the viscosity of a solvent represent transport properties that are directly related and affect the rates of mass transfer. In general, these properties are at least one order of magnitude higher and lower, respectively, compared to liquid solvents.[50] This means that the diffusion of a species through a medium at supercritical conditions can occur at a faster rate than in a liquid solvent, which implies that a solid will dissolve more rapidly in a SCF. In addition, a SCF will be

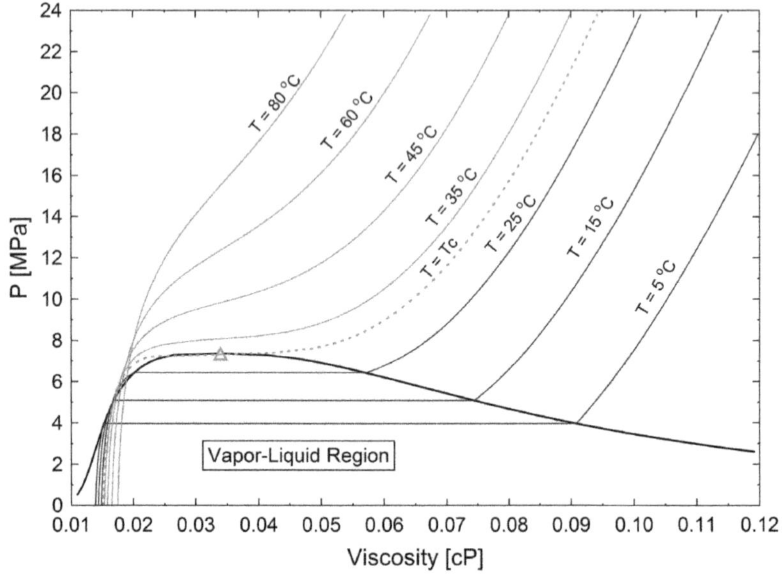

Figure 3.3 Pressure–viscosity (cinematic) diagram of CO_2 at different isotherms.

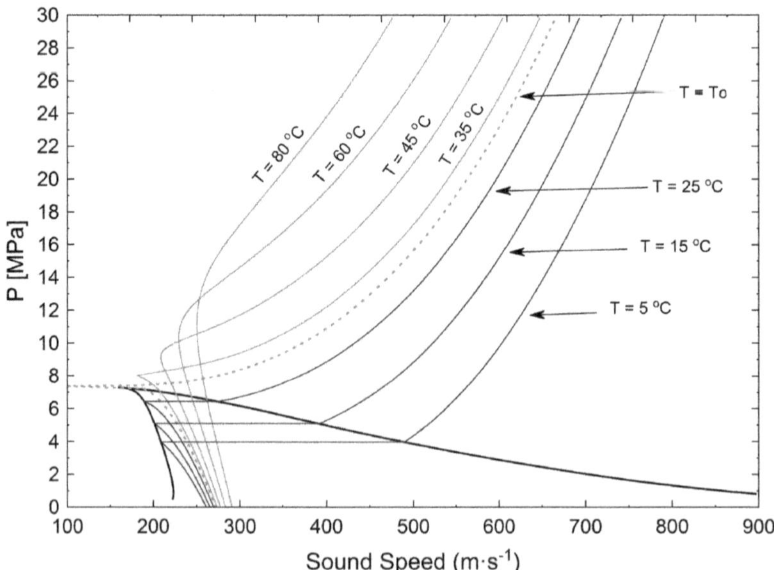

Figure 3.4 Pressure–sound speed diagram of CO_2 at different isotherms.

more efficient at penetrating a microporous solid structure.[50] Nonetheless, this does not mean that mass transfer limitations are not a limiting step in supercritical processes. In the pressure-viscosity diagram of Figure 3.3, it is clear that the thermodynamic behaviour of cinematic viscosity is similar to

Table 3.1 Solubility parameters for different organic solvents at normal conditions (25 °C and 1 atm).

Solvent	δ (MPa$^{1/2}$)
n-Pentane	14.4
n-Hexane	14.9
Chloroform	18.7
Acetone	19.7
Pyridine	21.7
Ethanol	26.2
Methanol	29.7
Water	48.0

that of the density of CO_2. Hence, it is possible to vary and tune a viscosity value by manipulating the pressure-temperature variables. However, both temperature and pressure have a dual and opposite effect when comparing density and viscosity values. At a constant supercritical temperature, the density of CO_2 can be increased by increasing the operating pressure to reach a high power solvent effect; however, an increase in pressure will also promote a considerable increase of the fluid viscosity.

As mentioned earlier, the solvent strength of a SCF can be adjusted or tuned by changing either temperature or pressure of the reaction/extraction system. For this purpose, the Hildebrand solubility parameter is a commonly applied concept for considering the application of liquid solvents in chemical processes. The Hildebrand solubility parameter is a measure of the cohesive energy density of a substance and it is quite useful for solvent selection. Table 3.1 presents the solubility parameter for some of the most common solvents.

Machida *et al.* presented a complete discussion about estimating the solubility parameter of SCFs including the use of equations of state (EoS) and models for its prediction.[52] According to these authors, differences in the solubility parameter must be within 4 MPa$^{1/2}$ on the basis of enthalpic interactions and at a solvent molar volume of 0.8 10^{-4} m^3 mol^{-1}. The molar volume of a SCF, however, is highly dependent on temperature and pressure and as the density of a SCF is increased, the 4 MPa$^{1/2}$ value can be made to increase as well. For the sake of comparison, the empirical correlation proposed by Giddings *et al.*[53] is here used for estimating the solubility parameter of SCF in SI units $\delta[\mathrm{MPa}^{1/2}] = 3.02 P_c^{1/2} \rho_r$, where P_c is in units of MPa and ρ_r is the reduced density ($\rho_r = \rho/\rho_c$).[53] Figure 3.5 presents the pressure–solubility parameter diagram as estimated using above equation and the reduced density values presented in Figure 3.1. This estimated and empirical diagram is in good agreement with the results presented by Machida *et al.*,[52] in which the Span and Wagner EoS[54] was used to calculate the solubility parameter of CO_2.[52] The appearance of the pressure–solubility parameter diagram of a pure component appears similar to its pressure–density diagram. The solubility parameter changes greatly with temperature or pressure along the critical isotherm and at the supercritical region,

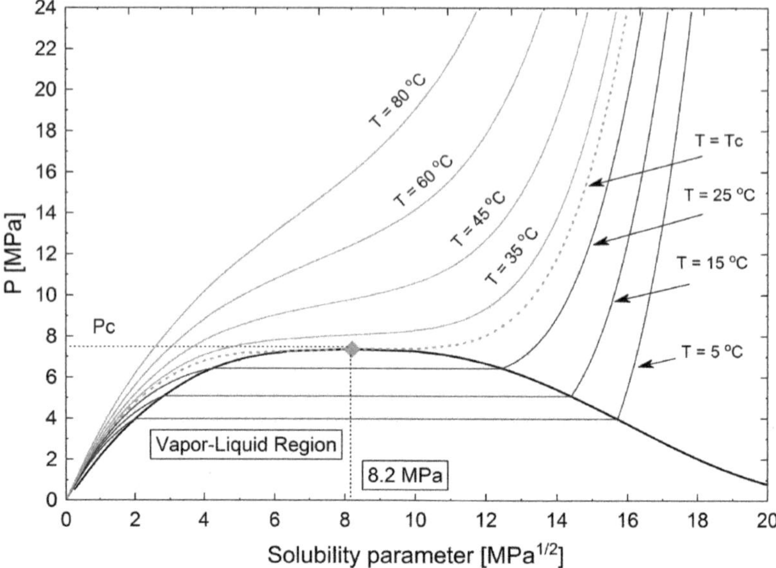

Figure 3.5 Solubility parameter of carbon dioxide as calculated by eqn (3.1).

following the same pressure–density–temperature behaviour of Figure 3.2. The value of the solubility parameter of CO_2 at its critical point (30.978 °C, 7.377 MPa, 467.6 kg m^{-3}) is 8.2 MPa$^{1/2}$ (Machida *et al.*[52]), while for water, this value is 17.1 MPa$^{1/2}$ at its critical point (373.946 °C, 22.064 MPa, 322 kg m^{-3}).[52]

Machida and co-workers also presented and discussed the solubility parameter of water at a wide range of pressure and temperature.[52] As shown by these authors, the solubility parameter of water can be manipulated in order to overlap those of many polar organic solvents, such as ethylene glycol (δ of 34.5 MPa$^{1/2}$), methanol (δ of 29.5 MPa$^{1/2}$), acetonitrile (δ of 24.1 MPa$^{1/2}$), acetone (δ of 19.8 MPa$^{1/2}$). However, the solubility parameter of CO_2 can be adjusted to overlap many nonpolar liquids such as *n*-hexane (δ of 14.9 MPa$^{1/2}$), toluene (δ of 18.3 MPa$^{1/2}$) and chlorodifluoromethane (δ of 17.3 MPa$^{1/2}$)[52] by changing their temperature and pressure conditions (Figure 3.5). Therefore, the highly tuneable solubility parameters of water and CO_2 in their supercritical state give them wide applicability as solvents.

The solubility parameter of a solvent can be modified by the addition of small amounts of co-solvents according to following equation,[52] $\delta_{mixture} = \phi_{solvent}\delta_{solvent} + \phi_{co\text{-}solvent}\delta_{co\text{-}solvent}$, where ϕ is the volume fraction and $\delta_{mixture}$ is the solubility parameter of the mixture. The ϕ can be estimated by assuming that the excess volume is equal to zero, which is a good approximation for small mole fractions of the co-solvent. If one uses a SCF as a solvent and a polar or associating substance as a co-solvent, $\delta_{mixture}$ can be changed by several units of MPa$^{1/2}$. Possibly, the $\phi_{co\text{-}solvent}\ \delta_{co\text{-}solvent}$ or $\phi_{co\text{-}solvent}\ \delta_{co\text{-}solvent}\ (\delta_{co\text{-}solvent} - \delta_{solvent})^2$ product may be a way to rank the effectiveness of co-solvents at a given set of conditions. The controllable solubility parameters of SCF also characterize their wide applicability when used as co-solvents. For example, CO_2 can be used as a co-solvent to modify the solubility parameter of liquid solvents (acetone, δ of 19.8 $MPa^{1/2}$; tetrahydrofuran, δ of 18.6 $MPa^{1/2}$) that are important in gas-expanded liquid applications. CO_2 is highly soluble in many ionic liquids ([C_4mim][Tf_2N], δ of 20.9 $MPa^{1/2}$; [C_4mim][PF6], δ of 22.7 $MPa^{1/2}$) and improvements in their solubility parameter are important in catalytic biphasic reactions.[52]

3.2.2 Dielectric Constant

The dielectric constant (ε) is closely related to important properties such as solubility and reaction rate constant. Regarding SCF, ε has a dependence on pressure similarly to that of density. Figure 3.6 presents the P *vs.* ε plot for pure CO_2 at different isotherms, including sub- and supercritical temperatures.

The ε values of pure CO_2 were estimated using the empirical equation proposed by Goldfarb *et al.* as a function of CO_2 density: $\varepsilon = 1 + 0.02260\rho + 1.23.10^{-4}\rho^{2}.^{55}$ For the calculations presented in Figure 3.6, the CO_2 density (in $dm^3\ mol^{-1}$) at different pressure–temperature conditions were obtained from the NIST webbook.[51]

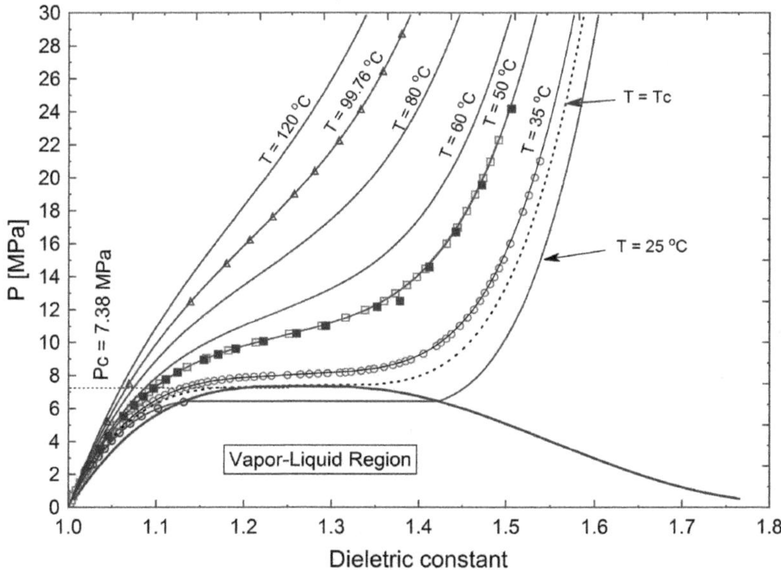

Figure 3.6 Dielectric constants of CO_2; symbols are experimental data ($T = 35$ °C and $T = 50$ °C are from Obriot and co-workers;[91] $T = 25$ °C, $T = 50$ °C and $T = 99.76$ °C from Michels and Kleerekoper[92]) and lines are estimated values based on following equation $\delta_{mixture} = \phi_{solvent}\delta_{solvent} + \phi_{co\text{-}solvent}\delta_{co\text{-}solvent}$.

The ε of $scCO_2$ presents small changes over a wide range of values of pressure and temperature, changing from 1.1 to 1.5 within the range used in this study (Figure 3.6), which is characteristic of a non-polar solvent. The ε of water, for example, exhibits an increase of more than 200% when the reduced pressure changes from 1.0 to 2.0.[50] Fluids whose ε varies significantly with the operating conditions have a great potential for applications as tuneable solvents. On the other hand, for systems whose ε is not affected by variations in the operating conditions (*P* and *T*), solvent tunability can be achieved by the addition of modifiers or co-solvents.

The effect of modifiers on the ε of SCF can be illustrated by the ethane–CO_2 system modified with methanol. Figure 3.7 shows that the ε values of CO_2 and ethane (non-polar solvent) increases with an increase in the methanol content of the mixture. The ε value for both systems approaches that of pure methanol at high methanol mole fractions and that of the pure compounds (CO_2 and ethane) at low methanol mole fractions. Both pressurized pure solvents, CO_2 and ethane, have similar pure-component dielectric properties, as observed in Figure 3.7 at zero methanol molar fraction. However, the ε value of the CO_2–methanol system is higher than that of the ethane–methanol system at any composition of the binary mixtures. This phenomenon suggests that the effect of modifiers on the

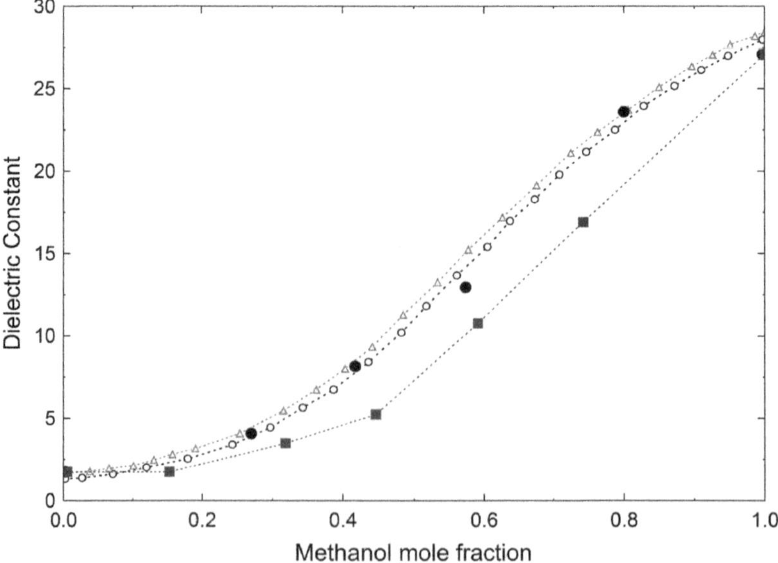

Figure 3.7 Dielectric constant of CO_2–methanol and ethane–methanol systems at different conditions: the experimental data (filled symbols) came from Anastas[50] and Roškar and co-workers (17.5 MPa and 65 °C),[56] while the dashed lines are connecting the points reproduced from Goldfarb and co-workers at 10 MPa and 52 °C (circles) and 20 MPa and 52 °C (triangles).[55]

dielectric properties of SCF depends on specific interactions between SCF and the co-solvent. In the CO_2–methanol system, the effect of methanol on the dielectric properties can be related to the formation of CO_2–methanol complexes and to CO_2–methanol intermolecular interactions resulting from the quadrupole moment of the CO_2 molecules. By contrast, ethane is not expected to exhibit either of these two properties.[50,56] Other parameters affecting the dielectric constant of supercritical systems containing a modifier/co-solvent are temperature and density. In the CO_2–methanol and ethane–methanol systems,[50] the dielectric constants increase with temperature reductions and density increments.[56]

3.2.3 Phase Equilibrium

The knowledge of the phase behaviour of binary and multicomponent systems plays an important role in process design and optimization and this is a major task for high-pressure systems based on SCF, in which the influence of pressure and temperature is rather complex. For example, multiple phases may be present in these systems, such as vapour–liquid–liquid or solid–vapour–liquid equilibria. In many cases, the operation of a process with SCF under multi-phase conditions may be undesirable and this can only be avoided if the overall phase behaviour is first investigated.[50] The limiting case of equilibrium between two components (binary systems) provides a convenient starting point in the understanding of multicomponent phase behaviours. In this section, a brief discussion is given about the phase behaviour for CO_2 with ethanol. However, the readers are encouraged to consult other literature for further information about the general subject of thermodynamics and phase equilibria at high-pressure systems.[50,57,58]

In Figure 3.8 and Figure 3.9, the continuous and dashed lines (saturated liquid and vapour lines, respectively) were calculated with the Peng-Robinson EoS with the van der Waals quadratic mixing rule (PR-vdW2), with binary interaction parameters k_{12} of 7.8332×10^{-2} and l_{12} of -3.0999×10^{-2}.[59]

The binary system $CO_2(1)+\text{ethanol}(2)$ presents a diagram of type I, following the Scott and van Konynenburg classification,[60] in which the critical line (critical locus) of the binary mixtures continuously connects the critical point of each of the pure components. This means that a vapour–liquid two-phase equilibrium exists at any point inside the region delimited by the vapour pressure curves of CO_2 and ethanol and below the critical line, where $V=L$. In mixtures that exhibit such behaviour, the vapour–liquid equilibrium envelop shifts to pressures above the critical pressures of both pure compounds. Thus, at a fixed temperature and global composition, the pressure needed to reach a homogenous phase at high pressures must be higher than the critical pressure of pure compounds; otherwise, vapour–liquid two-phases may coexist in the system. Therefore, it is crucial to know the phase behaviour of SCF + co-solvent (or an organic solvent as a modifier) systems to correctly define the process conditions of pressure and temperature.

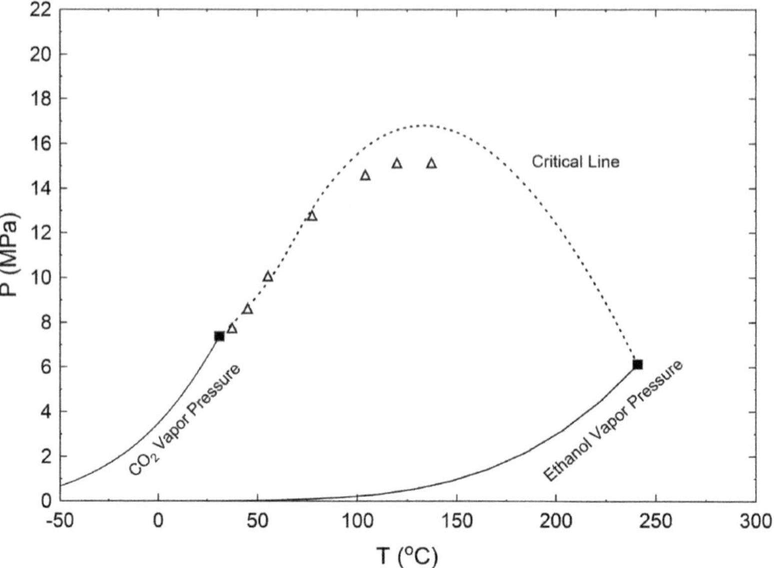

Figure 3.8 Critical point loci for $CO_2(1) + \text{ethanol}(2)$ binary mixtures, with the experimental data (triangles) coming from Yeo and co-workers[93] and the critical line calculated using the PR-vdW2 model.

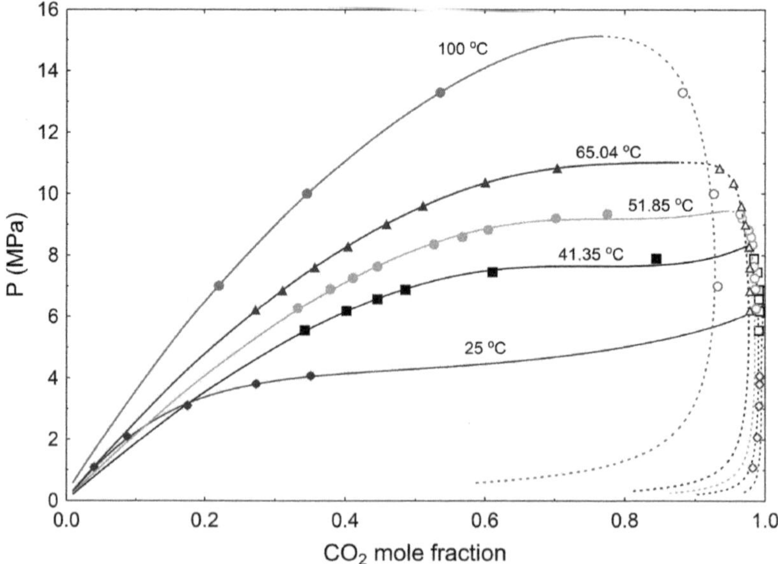

Figure 3.9 Pressure-composition diagram for $CO_2(1) + \text{ethanol}(2)$ at different temperatures; continuous (saturated liquid) and dashed (saturated vapour) lines are calculated with the thermodynamic model (PR-vdW2). Filled symbols represent bubble points and open symbols represent dew points.

In Figure 3.9, pressure-composition isotherms are presented for the system $CO_2(1) + \text{ethanol}(2)$. It can be seen that, at given temperature and composition, the pressure needed to get the system into a liquid phase (region above and left of the continuous lines) is higher when the CO_2 composition is increased. Also, high temperatures result in high pressures of saturation. Thus, starting from a pure CO_2 condition and systematically adding ethanol (co-solvent) into the system, higher temperatures and pressures are needed to reach the supercritical condition of the mixture. After a maximum condition (maximum point in the critical line in Figure 3.8), the pressure needed to reach the critical condition is decreased as the temperature and the ethanol content in the mixture are increased.

The complexity of the observed phase diagrams clearly demonstrates that phase behaviour under high-pressure conditions (or systems containing a SCF) can vary markedly. This is particularly important for chemical processes based on SCF technology since the thermodynamic, phase equilibrium and properties of the complex mixture have a direct impact on the process performance. Knowledge of phase behaviour is therefore essential for the proper interpretation of experimental data and results derived from systems involving $scCO_2$.

3.3 Application of CO_2 for Biomass Pre-treatment and Fractionation

This section reviews some of the most relevant publications released so far about the use of CO_2 for biomass pre-treatment and fractionation, with focus on its ability to enhance the accessibility of glucans (starch and cellulose) to enzymatic hydrolysis and fermentation.

3.3.1 Use of $scCO_2$ Under Subcritical Water (CO_2–H_2O Mixtures)

Several authors have used CO_2-impregnation strategies for biomass pre-treatment processes. Impregnation with CO_2 generates carbonic acid (H_2CO_3) inside the biomass fibrous structure and its distribution in the fibre network depends on the biomass moisture content, which must be below fibre saturation to facilitate the permeation of the gaseous phase. Hence, with the use of a weak acid such as H_2CO_3, the formation of fermentation inhibitors (mostly furan compounds) is very low, hemicellulose sugars are recovered as high value-added oligomers and lignin condensation is avoided.

Gurgel and co-workers studied the pre-treatment of sugarcane bagasse using a mixture of H_2O–CO_2 at an initial CO_2 pressure of 6.8 MPa and a liquid-to-solid ratio of 12 : 1.[61] Experiments were carried out for 17.6 to 102.4 min in the range of 93.8 to 136.2 °C, in which the reactor inner pressure reached 8.62 to 12.96 MPa, respectively. The maximum recoveries of water-soluble xylan oligomers was 16.8 wt% at 100 °C for 30 min and 4.9 wt% at 115 °C for 60 min. The maximum total xylan recovery in pre-treatment hydrolysates (monomers and oligomers) was equivalent to 10.34 g L^{-1} at 100 °C for 30 min and 8.90 g L^{-1} at 115 °C for 60 min, while glucan recoveries in the pre-treated material ranged from 89.9% at 130 °C for 90 min to 100% at 100 °C for 30 min. Hence, lower temperatures and shorter times were able to extract a high proportion of cane bagasse xylans while higher temperatures and longer times reduced the amount of water-soluble oligomers but glucan recovery in pre-treated materials was partially compromized.

Morais and co-workers reviewed the CO_2-assisted hydrothermal hydrolysis of lignocellulosic materials to soluble sugars (mostly monosaccharides) for their subsequent conversion to biofuels, chemicals and biomaterials.[27] The solubility of CO_2 changed with temperature and pressure and its use resulted in significant improvements on the hydrolysis yields, compared to the auto-hydrolysis approach. CO_2 can affect the pH of the system, depending on the pressure applied to the medium; so, the solubility of the gas can be altered depending on the pre-treatment conditions, generating carbonic acid which, together with other organic acids that are released in the reaction environment, promotes swelling and partial acid hydrolysis of biomass macromolecular components. In the absence of CO_2, high temperatures are required for the release of hydronium ions from water and the *in situ* generation of carbonic acid improves the pre-treatment efficiency.[3] The hemicellulose fraction, being a less crystalline and well-organized macromolecular structure than cellulose, is by far more susceptible to acid hydrolysis. Also, it has been demonstrated that, for the pre-treatment of aspen wood and corn stover, pre-treatment severity was increased by the addition of CO_2 but, compared to auto-hydrolysis, lower concentrations of formic, levulinic, malonic and acetic acids (among others) were generated in the reaction medium, thus suggesting the occurrence of lower levels of sugars degradation. This way, the use of CO_2 enhanced the selectivity for hemicellulose hydrolysis, increased the concentration of xylose in acid hydrolysates and improved the recovery of glucans (mostly cellulose), which remained relatively intact in the resulting pre-treated substrate.[62] It is widely known that cellulose requires more drastic pre-treatment conditions to be partially hydrolyzed and this can generate much greater amounts of dehydration by-products from hemicellulose sugars (mostly furfural). With the addition of CO_2, pre-treatment temperatures are lower compared to auto-hydrolysis and higher sugar yields may be achieved because the whole process is less aggressive and lower concentrations of inhibitory compounds are accumulated in the pre-treatment water-solubles. Also, acid corrosion is lower compared to the use of strong mineral acids and neutralization is not required after pre-treatment.[29] At the end of the process, CO_2 can be recycled in the process and volatile extractive components can be recovered for further use.[3]

Silveira and co-workers reviewed the use of different types of pre-treatment for the production of cellulosic ethanol and, among these, the advantages of using CO_2 as an acid impregnating agent are described.[3] Such advantages include low cost, low corrosion, low toxicity and low operational risk. Besides, enzymatic hydrolysis can be carried out without neutralization of acid hydrolysates because carbonic acid is decomposed to CO_2 after the pre-treatment process. It is also remarkable that CO_2 is constantly formed in the cellulosic ethanol production process because of fermentation and this gas can be stored, pressurized and sent to the pre-treatment reactor. This way, this impregnation technology may have lower operational costs because CO_2 is generated *in situ* at the industrial site.

Since the use of CO_2–H_2O systems leads to the *in situ* formation of carbonic acid, van Walsum[63] proposed a combined severity factor for $scCO_2$ pre-treatment, which is based on the estimated pH value of the reaction environment at a given temperature and partial pressure of CO_2 as determined by the following equation: $CS_{P_{CO_2}} = \log\ (R_0) - (8.00\times10^{-6} \times T^2 + 0.00209 \times T - 0.216 \times \ln(P_{CO_2}) + 3.92$, where $CS_{P_{CO_2}}$ is the combined severity factor, R_0 is the severity factor (as previously described by Cuissinat and Navard[64]), T is the temperature (in °C) and P_{CO_2} is the partial pressure of CO_2 in the atmosphere.[63–65]

Ferreira-Leitão and co-workers studied the steam explosion of CO_2-impregnated sugarcane bagasse and sugarcane leaves using pre-treatment residence times of 5, 10 and 15 min and temperatures in the range of 190 °C to 220 °C.[33] Also, these results were compared with those derived from the use of SO_2 impregnation, primarily in relation to glucose and xylose yields that were obtained after a 96 h enzymatic hydrolysis with a mixture of Celluclast 1.5L and Novozym 188 at 2 wt% TS and 180 rpm. Under the same experimental conditions, CO_2-impregnation produced steam-exploded substrates with much lower susceptibility to enzymatic hydrolysis. However, under more drastic conditions of temperature and reaction time, higher yields were obtained with CO_2 compared to SO_2, with the advantage of releasing much lower levels of fermentation inhibitors such as furfural and HMF. Pre-treatment of SO_2-impregnated cane straw was best at 190 °C for 5 min, using 30 wt% TS plus 3 wt% SO_2 in relation to the moisture content of the raw material, producing substrates that resulted in glucose yields of 79.7 wt% after enzymatic hydrolysis. By contrast, steam explosion of CO_2-impregnated cane bagasse using identical experimental conditions yielded only 50.2 wt% glucose. However, by increasing the pre-treatment temperature to 205 °C and the reaction time to 10 and 15 min, higher glucose yields of 81.8 and 86.6 wt% were obtained after hydrolysis, respectively.

Corrales *et al.*, analyzed CO_2-impregnated steam-exploded cane bagasse by Electronic Transmission Microscopy (ETM), Scanning Electron Microscopy (SEM), X-ray Diffraction (XRD) and Infrared Spectroscopy (FTIR) to assess both chemical and structural changes that were associated with this pre-treatment technology.[66] The pre-treatment conditions applied in this study were those optimized previously by Ferreira-Leitão *et al.*[33] Both ETM and

SEM results showed that the structural modifications with CO_2 were less pronounced than those promoted by the use of SO_2 and this was related to the higher acidity of the latter impregnation strategy compared to the former. According to the XRD data, both impregnated steam-exploded samples had a higher crystallinity index compared to native bagasse (65.5 and 56.4% for SO_2 and CO_2 impregnation, compared to 48.0% of the starting material) and this was explained by the extent of hemicellulose removal that was achieved in each case.

The development of high value-added applications for lignin and lignin derivatives are often seen as efficient strategy to improve the viability of biorefineries that are based on biomass conversion technologies.[67] Lignin is usually recovered in biomass conversion processes by alkaline extraction and its recovery depends on a neutralization stage that is usually done by adding dilute mineral acids. The neutralizing power of CO_2 can also be used for this purpose and it has already been well documented that the quality of the lignin obtained in this way is much better than what is derived from conventional methods of neutralization. For instance, the Lignoboost® process was developed to improve the profitability of pulp and paper mills by increasing the recovery capacity of pulping chemicals and lignin from alkaline black liquors.[68] CO_2 was reported as the best neutralizing agent due to its favourable properties, yielding a cleaner lignin stream that is suitable for several applications in the chemical industry. This process was commercialized by Metso Corporation (Helsinki, Finland) and the first Lignoboost® commercial scale lignin production begun in February 2013 by Domtar Corporation at the Plymouth Mill in North Carolina (USA), with a targeted rate of 75 tons per day.

Park and co-workers developed a lime/CO_2 pre-treatment technology for the production of ethanol from rice straw.[69] In this process, the role of CO_2 was to neutralize the whole pre-treatment slurry to pH values around 6.0, producing calcium carbonate at non-inhibitory concentrations for both enzymatic hydrolysis and fermentation. This way, 19.1 g L^{-1} ethanol was produced by SSCF, which corresponded to 74% of the theoretical yield from both glucose and xylose.

3.3.2 Use of CO_2 Under Supercritical Conditions

Several fluids can be applied under subcritical and supercritical conditions. However, CO_2 is the most widely used due to its thermodynamic and physicochemical properties. This solvent, besides having a critical point under relatively mild conditions, is relatively cheap, readily available, nontoxic, non-flammable, easily removed from the extracted material and environmentally friendly.[70] In addition, when $scCO_2$ is used as a solvent in the extraction of thermolabile compounds such as antioxidants, most of their antioxidant activity is preserved if compared to the use of organic solvents.[71,72] This is because in the conventional extraction process, such compounds are partially oxidized during the organic solvent recovery.

As mentioned earlier, supercritical or pressurized solvents are of particular interest for use in biomass pre-treatment since their physicochemical properties (density, diffusivity, viscosity, and dielectric constant) can be tuned by altering the pressure and temperature, allowing control of the solvating power and selectivity of the solvent during the extraction/reaction process.[73] In general, the use of CO_2 in biomass pre-treatment methods has been based on its ability to promote a better impregnation of the fibrous network, increasing the penetrability of chemical agents within the structure of the lignocellulosic material.[33] For instance, the use of $scCO_2$ has received attention for biomass pre-treatment due to the mild temperatures that can be employed, hence resulting in lower production of undesirable by-products mostly arisen from carbohydrate degradation. Also, the pre-treated material is recovered without any solvent residues because CO_2 is gaseous under normal environmental conditions.[74,75]

$scCO_2$ can diffuse through interspaces like gas and dissolve materials like liquid. Besides, the size of the CO_2 molecule is comparable to H_2O and NH_3. Explosive release of CO_2 pressure consequently increases the accessible surface area of the lignocellulosic materials. Furthermore, CO_2 can form carbonic acid in the presence of moisture, which favours the hydrolysis of plant cell wall macromolecular components such as polysaccharides and lignin.[76,77]

According to Sun and co-workers, the moisture content present in lignocellulosic materials is an important factor that significantly influences the enzymatic digestibility of $scCO_2$ pre-treated lignocellulosic materials.[77] Lv *et al.* reported that during the $scCO_2$ pre-treatment the addition of water–ethanol as co-solvents increases the lignin removal and further improves the enzymatic hydrolysis of corn stover.[78] Additionally, because cellulose is relatively stable in $scCO_2$ at 35 °C, simultaneous $scCO_2$ pre-treatment and enzymatic hydrolysis in one step has been applied to produce fermentable sugars.[79] It was also reported by these authors that simultaneous $scCO_2$ pre-treatment and enzymatic hydrolysis resulted in hydrolysis rates and glucose yields higher than discrete $scCO_2$ pre-treatment and enzymatic hydrolysis alone.

Benazzi and co-workers combined $scCO_2$ and ultrasound to enhance the enzymatic hydrolysis of sugarcane bagasse.[80] Compared to native bagasse, pre-treatment by $scCO_2$ increased the release of fermentable sugars by 280%, leading to glucan hydrolysis efficiencies as high as 74.2%. However, hydrolysis was carried out at high enzyme loadings and low TS of only 1 wt%. $scCO_2$ performed even better in the presence of ultrasound. Compared to $scCO_2$ alone, the combined pre-treatment increased by 16% the amount of fermentable sugar that was obtained by enzymatic hydrolysis. From these results, the ultrasound-assisted $scCO_2$ pre-treatment was claimed to be a promising alternative for improving the susceptibility of sugarcane bagasse to enzymatic hydrolysis using relatively low reaction temperatures and without the use of any hazardous solvents and/or catalysts.

Luterbacher and co-workers applied $scCO_2$ for the pre-treatment of several lignocellulosic materials using reactions times from 1 to 60 min and temperatures from 150 °C to 200 °C.[81] Mixed hardwoods, switchgrass and mixed perennial grasses (switchgrass and big blue-stem) were used at 20 and 40 wt% TS in the case of mixed hardwoods and at 40 wt% TS for the other materials. Significant yields of furfural and HMF were obtained, depending on the conditions used for pre-treatment. Because of the high content of hemicellulose in the pre-treated substrates, hydrolyses had to be performed using a mixture of cellulases (Spezyme CP®, Genencor), xylanases (Multifect®, Genencor) and β-glucosidases (Novozym 188®, Novozymes), containing a total of 15 FPU (filter paper units) g^{-1}, 30 CBU (cellobiose units) g^{-1} and 30 mg protein g^{-1} of xylanases, always in relation to the cellulose content of the pre-treated substrate. Pre-treatment of mixed hardwoods at 170 °C for 60 min using 20 or 40 wt% TS gave the highest glucose yield of 77 and 81% after enzymatic hydrolysis, respectively, with less than 1–2% glucose conversion to HMF. However, dehydration of hemicellulose sugars to furfural was high under these same pre-treatment conditions (over 20%). Likewise, switchgrass had a similar performance at 40 wt%, releasing low HMF and high furfural concentrations at the same pre-treatment severity and reaching 81% glucan conversion by enzymatic hydrolysis under identical experimental conditions. By contrast, pre-treatment of mixed perennial grasses at 40 wt% resulted in lower glucose yields of 68%, with less than 1% glucose conversion to HMF and 14% pentose conversion to furfural under similar operational conditions.

Zheng and co-workers applied $scCO_2$ in the pre-treatment of microcrystalline cellulose (Avicel) to investigate its effect on the increase of substrate accessibility to cellulolytic enzymes.[82] Such attempt was carried out by a typical CO_2-explosion approach (an explosive release of the reactor contents), causing disruption of the cellulose structure and an increase in its surface area. As the main result, pre-treatment with $scCO_2$ led to a 50% increase in the rate of enzymatic hydrolysis.

Phan and Tan used $scCO_2$ followed by alkaline hydrogen peroxide (H_2O_2) under ultrasonic irradiation for the pre-treatment of sugarcane bagasse.[83] Samples of bagasse with 80% moisture content were impregnated with CO_2 at 20.6 MPa and submitted to three different temperatures, 393, 423 and 453 K, to achieve the supercritical state during 1 h. After that, the $scCO_2$-treated bagasse was treated with alkaline H_2O_2 under different concentrations, with and without process intensification by the use of ultrasonic irradiation. The $scCO_2/H_2O_2$ process produced highly accessible substrates for enzymatic hydrolysis. Glucose yields of 97.8% were obtained after 72 h of hydrolysis using 15 FPU g^{-1} cellulose, while alkaline H_2O_2 alone produced substrates with four times lower accessibility to hydrolysis. The ideal conditions for the combined $scCO_2/H_2O_2$ pre-treatment process were 460 K, 15.6 MPa and 40 min for the $scCO_2$ stage and 1 wt% H_2O_2 for the subsequent stage. A similar work with $scCO_2$ followed by ultrasound allowed the recovery of 350 g kg^{-1} of fermentable sugars from sugarcane bagasse.[80]

3.4 Use of Co-solvents in CO_2-based Pre-treatment Methods

As mentioned above, $scCO_2$ is attractive as a pre-treatment method for lignocellulosic materials due to its high diffusivity combined with its easily tuneable solvent strength.[84] However, $scCO_2$ has several polarity limitations and it is known that $scCO_2$ alone does not lead to pre-treated substrates with high susceptibilities to enzymatic hydrolysis.[85] For instance, $scCO_2$ pre-treatment corn stover and switchgrass resulted in small (10–12%) increases in glucose yields after enzymatic hydrolysis and the use of co-solvents was envisaged as an effective way to improve substrate accessibility by removing lignin while producing substrates with low moisture content in high yields.[46] Organic solvents (*e.g.* ethanol and 1-butanol) are commonly added to the CO_2-extracting system to circumvent these limitations since solvent polarity is critical for extracting polar solutes from complex matrices such as the intricate associative structure of the plant cell wall. Since CO_2 is gaseous at room temperature and pressure, only the extracted products remain after depressurization, causing very little harm to the extraction and integrity of thermolabile compounds.

Likewise other pre-treatment methods such as steam explosion, $scCO_2$ may lead to biopolymer degradation. At high pre-treatment severities, hemicelluloses are hydrolyzed to soluble sugars while lignin is chemically modified mainly by cleavage of aryl-ether bonds but condensation reactions involving lignin and carbohydrate by-products may also take place.[78] For instance, it has been demonstrated that, after $scCO_2$ pre-treatment of corn stover, many lignin droplets were observed on the surface of cellulose fibres and these collapsed lignin fragments are known to limit enzymatic hydrolysis by triggering hydrophobic interactions with the cellulolytic enzymes. Therefore, the migration of lignin to the liquid phase still remains as one of the key factors limiting pre-treatment performance with regard to substrate accessibility. For this reason, co-solvents have been used to extract the lignin component and produce better substrates for enzymatic hydrolysis.

Pasquini *et al.* investigated the use of ethanol–water and 1-butanol–water mixtures in the presence of $scCO_2$ for biomass delignification.[48,49] When using ethanol/water mixtures, the best results were reached under 16 MPa at 190 °C, leading to 43.7 and 4.9% of total mass recovery (pulp yield) and residual lignin content (93.1% delignification), while these numbers for sugarcane bagasse were 32.7 and 8.7%, respectively (88.4% delignification).[48] However, substrates with low lignin content were produced at the expenses of a huge carbohydrate mass loss. The application of a less polar solvent mixture (1-butanol–water) with $scCO_2$ led to higher levels of delignification.[49] By applying $scCO_2$ in the presence of 60% (vol vol^{-1}) 1-butanol in water, 94.5% delignification yields were reached but again with carbohydrate mass losses as high as 40%.

Kroon and co-workers demonstrated that the use of $scCO_2$ as an extraction co-solvent or as an anti-solvent for polysaccharide precipitation depends on the chemistry of the starting material as well as on applied experimental conditions involving temperature, pressure, time and ionic liquid ratio.[86] IL are very strong solvents and based on this work, $scCO_2$ may be an excellent anti-solvent for IL reaction systems. Barber and co-workers demonstrated that $scCO_2$ is a coagulation additive for chitin and cellulose and that it can be used for the energy-efficient recovery of the IL.[87] Such process is started by the chemisorption of CO_2 by the imidazolium IL ([C_2mim][OAc]) and this gradually decreases the solubility of chitin and cellulose.

Silveira *et al.* developed a pre-treatment method based on the IL-assisted supercritical organosolv ($scCO_2$/EtOH-IL) (ethanol-[C_4mim][OAc]-$scCO_2$) of sugarcane bagasse.[3] The effect of process conditions such as temperature, time, IL loading and operating pressure were investigated. However, sugarcane bagasse (1 mm particle size in average) was pre-extracted with ethanol to allow the evaluation of pre-treatment efficiency in the absence of low molar mass extractive components. Pre-treatment of sugarcane bagasse (2 g) with 40 g $scCO_2$ and 2 g [C_4mim][OAc] at 25 MPa led to its partial delignification but the resulting cellulosic substrate was highly accessible to hydrolysis, producing 70 and 90% of glucose and xylose yields after 12 h using 5 wt% TS and Cellic CTec2 (Novozymes) at 10 mg g^{-1} TS, respectively. The main effect of $scCO_2$/EtOH-IL was the delignification probably due to the ability of [C_4mim][OAc] to catalyze the cleavage of aryl-ether bonds and cause partial lignin condensation at high temperatures.[88] Hence, cane bagasse was made highly accessible to hydrolysis even without high levels of delignification, meaning that the residual lignin was not an impediment for the concerted action of Cellic CTec2 enzymes. However, as mentioned before, IL-based pre-treatments can only be applied at industrial scale if high IL recoveries are reached. For the $scCO_2$/EtOH-IL process, the operation pressure strongly regulated [C_4mim][OAc] (91 to 95%) recovery while polysaccharides could be recovered in high yields: 100% for anhydroglucose and 93 to 97% for anhydroxylose. This fact also indicates the effectiveness of $scCO_2$ as polysaccharide coagulation agent,[87] since lignin is solubilised in the IL phase and $scCO_2$ induces polysaccharides precipitation.

3.5 Scale-up of CO_2-based Methods for Biomass Pre-treatment

The recommended procedure for scaling up an $scCO_2$-based pre-treatment process follows a stage gate approach, which is similar to what is normally applied in the traditional Chemical Process Industry (CPI). However, for this specific application, several singularities and unique challenges must be taken in consideration. In general, the scaling up of a chemical process must be grounded on preliminary results from studies in which important early-stage tools for proving concepts were settled. These and other considerations are detailed in the following sub-sections.

3.5.1 CO_2 Supply

The development of scCO_2 biomass pre-treatment technologies for the production of liquid fuels would need a huge availability of this material (CO_2) in the required quantity and on time, considering that it is a critical element to process execution. Therefore, it will be indispensable to seek ways to ensure a strong CO_2 supplier in the market. However, if CO_2 produced at the mill in the current fermentation process can be used, it is important to keep in mind the quality of this product. At the mill, CO_2 usually brings together significant amounts of water and other volatile impurities that must be eliminated prior to its use in biomass pre-treatment.

3.5.2 CO_2 Pressurizing

The development of scCO_2 biomass pre-treatment technologies also requires CO_2 pressurisation to achieve its supercritical conditions. This demands the use of compressors that depending on the process requirements may present special needs, besides a very precise dimensioning. As a result, the pressurizing step leads to high energy consumption.

3.5.3 Unit Operations

Considering the scCO_2 explosion processing, the tank required for biomass impregnation needs to obey a very rigid safety rules, since the area where it will be located will probably be considered as a classified area, which demands special attention. Considering that the impregnation process will run under very high pressure, special needs as differentiated constructive material of the vessel and a very robust instrumentation will bring a huge impact to the CAPEX of the project.

3.5.4 Reaction Feeding Mode

As the CO_2 use arises as an opportunity for technical improvements on biomass pre-treatment process, it is important to install the reaction feeding mode close to the pre-treatment unit, taking into consideration that the challenge for feeding impregnation under very high pressure is still under discussion (considering cost-effective aspects) and it does not involve a trivial solution in large scale. Nowadays, this process will probably run in batch mode, which impacts on productivity and equipment downtime in comparison with continuous technologies. At the end of the process, it would be mandatory to develop the concept of closed systems in which CO_2 recovery is fully addressed.

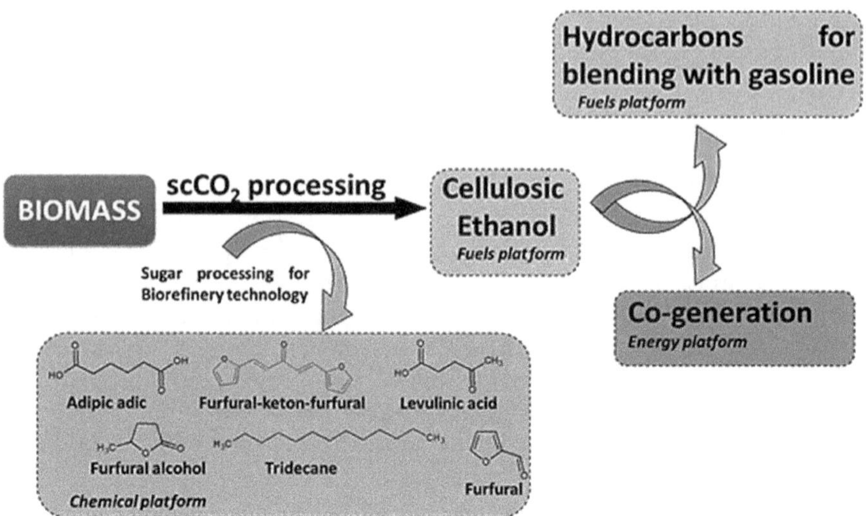

Figure 3.10 Representation of a biorefinery process based on $scCO_2$ to produce chemicals, fuels and energy from biomass.

From the scientific point of view, the application of $scCO_2$ is possible, but not feasible for industrial application yet, if the focus is only liquid fuels, due to the high cost required for building up a facility. However, for processes that are addressed to other high value-added platforms (*e.g.* chemical and energy) it might be an option.

Regarding the development of industrial process for biomass pre-treatment, recently, Kilambi and Kadam[89] deposited an application for a patent on behalf of Renmatix, for a pre-treatment technology using $scCO_2$ (120 °C and 25 MPa) followed by organosolv steps that combined use of C1–C5 alcohols. In general, the use of $scCO_2$ for biomass processing might lead to improvements in the biorefinery process with regard to the organic solvent recovery and to the extraction of added value chemicals as represented in Figure 3.10.

Biomass can be pre-treated by different technologies and in all of them, $scCO_2$ can be applied as auxiliary extraction solvent. As a result, liquid biofuels such as ethanol and butanol can be produced together with important chemicals for different industrial applications and, by doing so, both chemical and biochemical platforms are fully addressed. Residues that were not converted into fuels or chemicals might be burned for co-generation.

As a good example of this approach, da Costa Lopes *et al.* developed a sustainable process for the valorization of lignocellulosic materials by combining [C_2mim][OAc] with $scCO_2$ as the extraction co-solvent.[90] Polysaccharides were recovered from wheat straw in high yields and high levels of delignification were achieved with [C_2mim][OAc], from which a great deal of phenolic compounds could be obtained. The extraction of such phenolic compounds was carried out with Amberlite XAD-7 and the remaining [C_2mim][OAc] was recovered by $scCO_2$ in very high yields and purity.

3.6 Conclusions

The use of CO_2 in biomass pre-treatment has been demonstrated as a feasible technique with great potential for industrial applications. Hence, the knowledge about phase behaviour of systems involving CO_2 and organic co-solvents or modifiers has a central role in the process development and optimization. More studies about the combination of organic solvents (or other modifiers and catalysts) with $scCO_2$, ideally in a single pre-treatment step, might be a promising way to facilitate the deployment of this technology to an industrial scale. SCF technologies using CO_2 are also promising due to their economic and environmental performance. However, several bottlenecks must still be overcome such as improvements in the residual energy use to reduce costs. One of such bottlenecks is the development of methods for utilizing $scCO_2$ as a green solvent for continuous biomass processing. Also, there is a need to apply this technology to other lignocellulosic materials with different chemical composition and properties. Only this would be able to unveil the actual possibilities of this technology for the development of sustainable biorefinery processes.

Acknowledgements

This work was partially supported by CNPq (grants: 551404/2010-8; 311554/2011-3) and Araucaria Foundation (grants 3572013-35356 and 2742013-35145) (Brazil). M.H.L.S. and F.M.L.M. are also grateful to CAPES (Brazil) for providing scholarships to carry out their graduate studies.

References

1. S. H. Mood, A. H. Golfeshan, M. Tabatabaei, G. S. Jouzani, G. H. Najafi, M. Gholami and M. Ardjmand, *Renewable Sustainable Energy Rev.*, 2013, **27**, 77–93.
2. Y. Sun and J. Cheng, *Bioresour. Technol.*, 2002, **83**, 1–11.
3. M. H. L. Silveira, B. A. Vanelli, M. L. Corazza and L. P. Ramos, *Bioresour. Technol.*, 2015, **192**, 389–396.
4. T. L. Ogeda and D. F. Petri, *Quim. Nova*, 2010, **33**, 1549–1558.
5. V. Santos, R. Ely, A. Szklo and A. Magrini, *Renewable Sustainable Energy Rev.*, 2016, **53**, 1443–1458.
6. M. Balat, *Energy Convers. Manage.*, 2011, **52**, 858–875.
7. T. C. Hsu, G. L. Guo, W. H. Chen and W. S. Hwang, *Bioresour. Technol.*, 2010, **101**, 4907–4913.
8. L. P. Ramos, *Quim. Nova*, 2003, **26**, 863–871.
9. X. Zhang, W. Qin, M. G. Paice and J. N. Saddler, *Bioresour. Technol.*, 2009, **100**, 5890–5897.
10. R. Maryana, D. Ma'rifatun, A. Wheni, K. Satriyo and W. A. Rizal, *Energy Procedia*, 2014, **47**, 250–254.
11. C. Martin, M. Marcet and A. B. Thomsen, *BioResources*, 2008, **3**, 670–683.

12. S. Zhao, M. Cheng, J. Li, J. Tian and X. Wang, *Chem. Commun.*, 2011, **47**, 2176–2178.
13. D. Gao, S. P. Chundawat, C. Krishnan, V. Balan and B. E. Dale, *Bioresour. Technol.*, 2010, **101**, 2770–2781.
14. Q. Yu, X. Zhuang, S. Lv, M. He, Y. Zhang, Z. Yuan, W. Qi, Q. Wang, W. Wang and X. Tan, *Bioresour. Technol.*, 2013, **129**, 592–598.
15. M. Sasaki, K. Goto, K. Tajima, T. Adschiri and K. Arai, *Green Chem.*, 2002, **4**, 285–287.
16. A. Hideno, H. Inoue, K. Tsukahara, S. Fujimoto, T. Minowa, S. Inoue, T. Endo and S. Sawayama, *Bioresour. Technol.*, 2009, **100**, 2706–2711.
17. R. Travaini, M. D. M. Otero, M. Coca, R. Da-Silva and S. Bolado, *Bioresour. Technol.*, 2013, **133**, 332–339.
18. L. W. Yoon, G. C. Ngoh, A. S. M. Chua and M. A. Hashim, *J. Chem. Technol. Biotechnol.*, 2011, **86**, 1342–1348.
19. A. M. da Costa Lopes, K. G. João, A. R. C. Morais, E. Bogel-Lukasik and R. Bogel-Lukasik, *Sustainable Chem. Process.*, 2013, **1**, 3.
20. L. Mesa, E. González, C. Cara, E. Ruiz, E. Castro and S. I. Mussatto, *J. Chem. Technol. Biotechnol.*, 2010, **85**, 1092–1098.
21. R. Sindhu, P. Binod, K. Satyanagalakshmi, K. U. Janu, K. V. Sajna, N. Kurien, R. K. Sukumaran and A. Pandey, *Appl. Biochem. Biotechnol.*, 2010, **162**, 2313–2323.
22. L. Mesa, E. González, E. Ruiz, I. Romero, C. Cara, F. Felissia and E. Castro, *Appl. Energy*, 2010, **87**, 109–114.
23. J. Yu, J. Zhang, J. He, Z. Liu and Z. Yu, *Bioresour. Technol.*, 2009, **100**, 903–908.
24. R. Velmurugan and K. Muthukumar, *Bioresour. Technol.*, 2011, **102**, 7119–7123.
25. P. Binod, K. Satyanagalakshmi, R. Sindhu, K. U. Janu, R. K. Sukumaran and A. Pandey, *Renewable Energ.*, 2012, **37**, 109–116.
26. L. D. Serna, C. O. Alzate and C. C. Alzate, *Bioresour. Technol.*, 2016, **199**, 113–120.
27. A. R. C. Morais, A. M. da Costa Lopes and R. Bogel-Lukasik, *Chem. Rev.*, 2015, **115**, 3–27.
28. E. R. Gouveia, R. T. d. Nascimento, A. M. Souto-Maior and G. J. d. M. Rocha, *Quim. Nova*, 2009, **32**, 1500–1503.
29. F. Carvalheiro, L. C. Duarte and F. M. Girio, *J. Sci. Ind. Res.*, 2008, **67**, 849–864.
30. A. S. da Silva, H. Inoue, T. Endo, S. Yano and E. P. S. Bon, *Bioresour. Technol.*, 2010, **101**, 7402–7409.
31. A. García, C. Cara, M. Moya, J. Rapado, J. Puls, E. Castro and C. Martín, *Ind. Crops. Prod.*, 2014, **53**, 148–153.
32. C. Carrasco, H. M. Baudel, J. Sendelius, T. Modig, C. Roslander, M. Galbe, B. Hahn-Hägerdal, G. Zacchi and G. Lidén, *Enzyme Microb. Technol.*, 2010, **46**, 64–73.
33. V. Ferreira-Leitão, C. C. Perrone, J. Rodrigues, A. P. M. Franke, S. Macrelli and G. Zacchi, *Biotechnol. Biofuels*, 2010, **3**, 7.

34. T. H. Kim and Y. Y. Lee, *Bioresour. Technol.*, 2005, **96**, 2007–2013.
35. J. B. Binder and R. T. Raines, *J. Am. Chem. Soc.*, 2009, **131**, 1979–1985.
36. A. Emmel, A. L. Mathias, F. Wypych and L. P. Ramos, *Bioresour. Technol.*, 2003, **86**, 105–115.
37. A. P. Pitarelo, C. S. da Fonseca, L. M. Chiarello, F. M. Gírio and L. P. Ramos, *J. Braz. Chem. Soc.*, 2016, **27**, 1889–1898.
38. S. Gámez, J. J. González-Cabriales, J. A. Ramírez, G. Garrote and M. Vázquez, *J. Food Eng.*, 2006, **74**, 78–88.
39. C. C. Geddes, M. Mullinnix, I. Nieves, J. Peterson, R. Hoffman, S. York, L. Yomano, E. Miller, K. Shanmugam and L. Ingram, *Bioresour. Technol.*, 2011, **102**, 2702–2711.
40. C. Geddes, J. Peterson, C. Roslander, G. Zacchi, M. Mullinnix, K. Shanmugam and L. Ingram, *Bioresour. Technol.*, 2010, **101**, 1851–1857.
41. C. Vanderghem, Y. Brostaux, N. Jacquet, C. Blecker and M. Paquot, *Ind. Crops. Product.*, 2012, **35**, 280–286.
42. D. Scordia, S. L. Cosentino, J. W. Lee and T. W. Jeffries, *Biomass Bioenergy*, 2011, **35**, 3018–3024.
43. G. J. de Moraes Rocha, C. Martin, I. B. Soares, A. M. S. Maior, H. M. Baudel and C. A. M. De Abreu, *Biomass Bioenergy*, 2011, **35**, 663–670.
44. J. Bludworth and F. C. Knopf, *J. Supercrit. Fluids.*, 1993, **6**, 249–254.
45. T. Reyes, S. Bandyopadhyay and B. McCoy, *J. Supercrit. Fluids.*, 1989, **2**, 80–84.
46. N. Narayanaswamy, A. Faik, D. J. Goetz and T. Y. Gu, *Bioresour. Technol.*, 2011, **102**, 6995–7000.
47. H. S. Lu, M. M. Ren, M. H. Zhang and Y. Chen, *Chin. J. Chem. Eng.*, 2013, **21**, 551–557.
48. D. Pasquini, M. T. B. Pimenta, L. H. Ferreira and A. A. S. Curvelo, *J. Supercrit. Fluids*, 2005, **34**, 125–131.
49. D. Pasquini, M. T. B. Pimenta, L. H. Ferreira and A. A. S. Curvelo, *J. Supercrit. Fluids..*, 2005, **36**, 31–39.
50. P. T. Anastas, W. Leitner and P. G. Jessop, Handbook of Green Chemistry, *Green Solvents, Supercritical Fluids*, Wiley-VCH, 2013, vol 4.
51. P. J. Linstrom and W. G. Mallard, Book of the NIST Chemistry Web, http://webbook.nist.gov/chemistry/fluid, 2016.
52. H. Machida, M. Takesue and R. L. Smith, *J. Supercrit. Fluids.*, 2011, **60**, 2–15.
53. J. C. Giddings, M. N. Myers and J. W. King, *J. Chromatogr. Sci.*, 1969, **7**, 276–283.
54. R. Span and W. Wagner, *J. Phys. Chem. Ref. Data*, 1996, **25**, 1509–1596.
55. D. L. Goldfarb, D. P. Fernández and H. R. Corti, *Fluid Phase Equilib.*, 1999, **158**, 1011–1019.
56. V. Roškar, R. A. Dombro, G. A. Prentice, C. R. Westgate and M. A. McHugh, *Fluid Phase Equilib.*, 1992, **77**, 241–259.
57. M. McHugh and V. Krukonis, *Supercritical Fluid Extraction–Principles and Applications*, Butterworth-Heineman, Boston, MA, USA, 1994.

58. J. M. Prausnitz, R. N. Lichtenthaler and E. G. de Azevedo, *Molecular Thermodynamics of Fluid-phase Equilibria*, Pearson Education, Upper Sadler River, NJ, USA, 1998.
59. O. A. Araújo, F. R. Silva, L. P. Ramos, M. K. Lenzi, P. M. Ndiaye and M. L. Corazza, *J. Chem. Thermodyn.*, 2012, **47**, 412–419.
60. R. L. Scott and P. H. van Konynenburg, *Discuss. Faraday Soc.*, 1970, **49**, 87–97.
61. L. V. A. Gurgel, M. T. B. Pimenta and A. A. da Silva Curvelo, *Ind. Crops. Product.*, 2014, **57**, 141–149.
62. G. P. Van Walsum, M. Garcia-Gil, S.-F. Chen and K. Chambliss, *Appl. Biochem. Biotechnol.*, 2007, 301–311.
63. G. P. van Walsum, *Appl. Biochem. Biotechnol.*, 2001, **91–93**, 317–329.
64. C. Cuissinat and P. Navard, Swelling and Dissolution of Cellulose Part 1: Free Floating Cotton and Wood Fibres in N-Methylmorpholine-N-oxide–Water Mixtures, in *Macromol. Symp.*, 2006.
65. G. P. van Walsum and H. Shi, *Bioresour. Technol.*, 2004, **93**, 217–226.
66. R. C. N. R. Corrales, F. M. T. Mendes, C. C. Perrone, C. Sant'Anna, W. de Souza, Y. Abud, E. P. D. Bon and V. Ferreira-Leitao, *Biotechnol. Biofuels*, 2012, **5**, 36.
67. A. J. Ragauskas, G. T. Beckham, M. J. Biddy, R. Chandra, F. Chen, M. F. Davis, B. H. Davison, R. A. Dixon, P. Gilna and M. Keller, *Science*, 2014, **344**, 1246843.
68. P. Tomani, *Cellul. Chem. Technol.*, 2010, **44**, 53.
69. J.-y. Park, R. Shiroma, M. I. Al-Haq, Y. Zhang, M. Ike, Y. Arai-Sanoh, A. Ida, M. Kondo and K. Tokuyasu, *Bioresource Technol.*, 2010, **101**, 6805–6811.
70. L. T. Danh, R. Mammucari, P. Truong and N. Foster, *Chem. Eng. J.*, 2009, **155**, 617–626.
71. L. Casas, C. Mantell, M. Rodríguez, A. Torres, F. Macías and E. M. de la Ossa, *Chem. Eng. J.*, 2009, **152**, 301–306.
72. S. López-Sebastián, E. Ramos, E. Ibánez, J. M. Bueno, L. Ballester, J. Tabera and G. Reglero, *J. Agric. Food Chem.*, 1998, **46**, 13–19.
73. C. Pronyk and G. Mazza, *J. Food Eng.*, 2009, **95**, 215–226.
74. K. H. Kim and J. Hong, *Bioresource Technol.*, 2001, **77**, 139–144.
75. A. L. F. Santos, K. Y. F. Kawase and G. L. V. Coelho, *J. Supercrit. Fluids*, 2011, **56**, 277–282.
76. P. Alvira, E. Tomas-Pejo, M. Ballesteros and M. J. Negro, *Bioresour. Technol.*, 2010, **101**, 4851–4861.
77. S. Sun, S. Sun, X. Cao and R. Sun, *Bioresour. Technol.*, 2016, **199**, 49–58.
78. H. S. Lv, L. Yan, M. H. Zhang, Z. F. Geng, M. M. Ren and Y. P. Sun, *Chem. Eng. Technol.*, 2013, **36**, 1899–1906.
79. G. Muratov and C. Kim, *Biotechnol. Bioprocess Eng.*, 2002, **7**, 85–88.
80. T. Benazzi, S. Calgaroto, V. Astolfi, C. Dalla Rosa, J. V. Oliveira and M. A. Mazutti, *Enzyme Microb. Technol.*, 2013, **52**, 247–250.
81. J. S. Luterbacher, J. W. Tester and L. P. Walker, *Biotechnol. Bioeng.*, 2010, **107**, 451–460.

82. Y. Zheng, H.-M. Lin, J. Wen, N. Cao, X. Yu and G. T. Tsao, *Biotechnol. Lett.*, 1995, **17**, 845–850.
83. D. T. Phan and C.-S. Tan, *Bioresour. Technol.*, 2014, **167**, 192–197.
84. M. Herrero, J. A. Mendiola, A. Cifuentes and E. Ibáñez, *J. Chromatogr. A*, 2010, **1217**, 2495–2511.
85. M. A. Gao, F. Xu, S. R. Li, X. C. Ji, S. F. Chen and D. Q. Zhang, *Biosyst. Eng.*, 2010, **106**, 470–475.
86. M. C. Kroon, J. van Spronsen, C. J. Peters, R. A. Sheldon and G.-J. Witkamp, *Green Chem.*, 2006, **8**, 246–249.
87. P. S. Barber, C. S. Griggs, G. Gurau, Z. Liu, S. Li, Z. Li, X. Lu, S. Zhang and R. D. Rogers, *Angew. Chem.*, 2013, **125**, 12576–12579.
88. X. Sun, C. Huang, Z. Xue, C. Yan and T. Mu, *Energ. Fuel.*, 2015, **29**, 2564–2570.
89. S. Kilambi and K. L. Kadam, US8968479 B2, 2015.
90. A. M. da Costa Lopes, M. Brenner, P. Fale, L. B. Roseiro and R. Bogel-Lukasik, *ACS Sustainable Chem. Eng.*, 2016, **4**, 3357–3367.
91. J. Obriot, J. Ge, T. Bose and J.-M. St-Arnaud, *Fluid Phase Equilib.*, 1993, **86**, 314–350.
92. A. Michels and L. Kleerekoper, *Physica*, 1939, **6**, 586–590.
93. S.-D. Yeo, S.-J. Park, J.-W. Kim and J.-C. Kim, *J. Chem. Eng. Data*, 2000, **45**, 932–935.

CHAPTER 4

Enzyme-based Biomass Catalyzed Reactions in Supercritical CO_2

MAJA LEITGEB,* KATJA VASIĆ AND ŽELJKO KNEZ

University of Maribor, Faculty of Chemistry and Chemical Engineering, Laboratory for Separation Processes and Product Design, Smetanova ulica 17, SI-2000 Maribor, Slovenia
*Email: maja.leitgeb@um.si

4.1 Introduction

With the increasing and undoubtedly necessary environmental protection, a reduction of the use of organic solvents in chemical processes is inevitable. Organic solvents represent a major class of air pollutants, as they are volatile organic compounds. Supercritical carbon dioxide ($scCO_2$) as such is a green solvent and is often discussed as an alternative, since it is an environmentally benign reaction medium, due to its non-toxic and non-flammable nature. As CO_2 is a by-product of industrial processes, it is inexpensive and climate neutral and has a low critical temperature (31.1 °C) and pressure (7.38 MPa).[1,2] $ScCO_2$ differs from ordinary solvents in having both liquid-like high solubility and density and gas-like high diffusivities and low viscosities, in which all these properties can easily be controlled by the manipulation of temperature and pressure.[3,4] Near the critical point, significant changes in density and density-dependant properties can be achieved by applying small changes in temperature or pressure.[4–6]

Green Chemistry Series No. 48
High Pressure Technologies in Biomass Conversion
Edited by Rafał M. Łukasik

Published by the Royal Society of Chemistry, www.rsc.org

A lot of interest on research about enzymatic reactions in $scCO_2$ started in the early 80's. This medium is very interesting since it is environmentally benign and the advantages of using it include the total replacement of organic solvent, coupled with higher diffusivity and lower viscosity that reduce interphase transport limitations, enhance the reaction kinetics, and tune reaction selectivity due to an appreciable increase in the local concentration of substrate and catalyst. Its tunable solvating power facilitates the separation of reactants, products, and catalysts after reaction and, hence, the integration of biocatalytic and downstream processing steps in a single bioreactor.[7] Enzymatic esterification, transesterification, interesterification and oil hydrolysis reactions have been predominantly studied.

4.2 Enzymatic Reactions in $scCO_2$

Oil compounds seem to be suitable for transformation in this medium since $scCO_2$ is an excellent solvent for hydrophobic compounds and lipases are quite robust enzymes and have excellent stability and residual activity in this medium even though at some experimental conditions conformational changes and structural alterations may occur.

As mentioned before enzymatic transformation of oil compounds, such as hydrolysis or esterification in $scCO_2$ were frequently studied in high-pressure batch reactors (Figure 4.1). Esterified or hydrolyzed products obtained this way might find applications in food and personal care products. A good

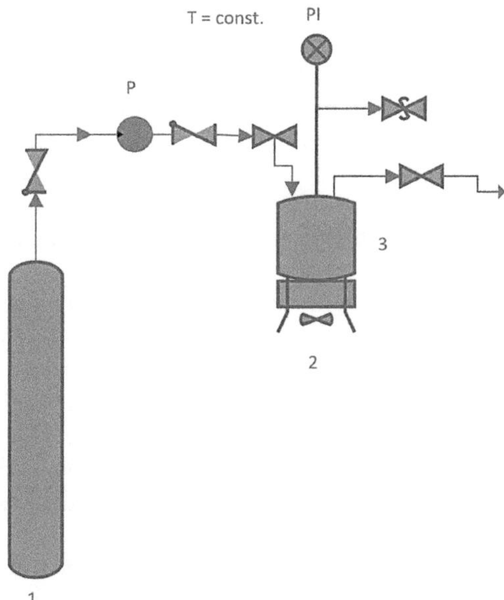

Figure 4.1 Schematic diagram of high pressure device: 1 – CO_2 tank, 2 – magnetic stirrer and heater, 3 – high pressure batch reactor, P – pump, PI – pump indicator.

example is enzymatic hydrolysis of conjugated linoleic acid (CLA)-enriched anhydrous milk fat using $scCO_2$, which was shown to be an efficient method to obtain free fatty acid (FFA), reaching a maximum of $86.79 \pm 7.28\%$ (w/w) in 2 h at 55 °C, 23 MPa and a fat–water ratio of 1 : 5 (mol/mol), when commercial enzyme Lipozyme TL IM was utilized as a catalyst.[8] CLA has beneficial health effects, such as anticancer functioning, prevention of cardiovascular diseases, and positive immune and inflammatory responses. Another study shows that $scCO_2$ has been studied as a medium for esterification of camel hump fat and tristearin in producing cocoa butter analog using immobilized lipase Lipozyme TL IM as a biocatalyst.[9] Lipase-catalyzed synthesis of terpene esters which may be applied as fragrances or flavours[10] and lipase-catalyzed synthesis of sugar fatty acid esters which possess good antimicrobial activity[11] were performed in high yields in $scCO_2$, as well.

Since high-pressure equipment is connected to high costs, compounds with added value shall be synthesized using $scCO_2$, such as synthesis of novel structured phenolic lipids which have potential biological properties, such as antioxidant, antiviral, anti-inflammatory and UV absorbing properties.[12,13] Transesterification enables the improvement through the solubility of phenolic acids in lipophilic media which have potential applications in food processing, pharmaceutical and cosmetic industries.[14–16] Enzymatic synthesis of phenolic lipids using flaxseed oil and ferulic acid was successfully performed in a $scCO_2$ batch reactor with very high yield of $58.0 \pm 2\%$.[17]

Even though there are many advantages of performing enzymatic reactions using CO_2 in the single supercritical state as a solvent the major drawback is the high pressure (on the order of hundreds of bars) required to ensure entire solubility of many organic compounds in CO_2. Therefore, advantages of using CO_2 expanded reaction media (*i.e.* reaction media where $scCO_2$ is used as a solvent although the reaction is carried out in the liquid reaction bulk in subcritical conditions) were applied in many cases. So the esterification of oleic acid by 1-octanol over Lipozyme RM IM in CO_2 expanded reaction mixture resulted in enhancing the reaction kinetics respect to the solvent-free system, requiring lower enzyme concentrations desired at commercial industrial scales.[7]

To overcome the problem with before mentioned problem with high pressure demand due to limited solubility of compounds another solution is the use of co-solvents, where some organic solvents or ionic liquids (ILs) may be used. Enzymatic synthesis in $scCO_2$/organic solvent and $scCO_2$/IL media proved to be an affective preparative method for the synthesis of terpene esters applicable in food industry.[18] Figure 4.2 shows variations in solubility of R,S-1-phenylethanol/vinyl acetate/$scCO_2$ system at different temperatures and pressures.

ILs in combination with $scCO_2$ are a useful reaction media. Clean chemical processes providing directly pure products results by using appropriate enzyme-ILs systems in combination with SCFs contain an exciting potential for developing the upcoming green chemical industry.[19] Using a combination of these two solvents may improve the enzyme stability and the

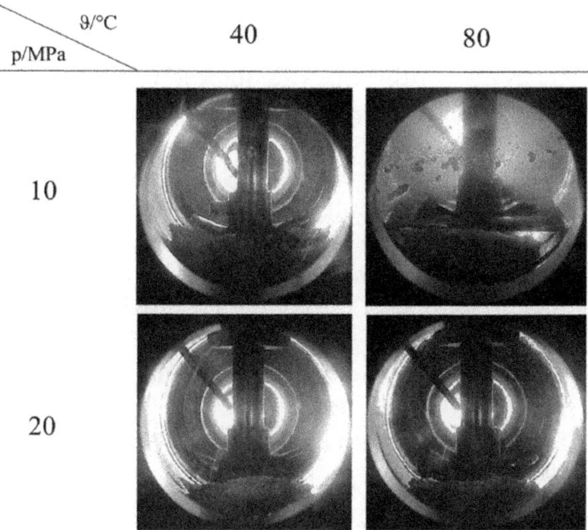

Figure 4.2 Phase behaviour of (R,S)-1-phenylethanol–vinyl acetate–$scCO_2$ system at different temperatures and pressures.

production of less or no pollutants. Very high efficiency at esterification of *n*-butanol and D,L-lactic acid (D, L-LA), catalyzed by *Candida antarctica* lipase B (Novozym 435) in $scCO_2$/IL (CYPHOS IL-201) medium was achieved which is one of many confirmations of previous statement.[20]

$ScCO_2$ was also found as a good pre-treatment agent before an enzyme was used for biocatalysis. This is very useful for industrial applications since high activity of an enzyme can be reached with this method. FT-IR analyses confirmed some structural changes in the secondary structure of *Candida rugosa lipase* (CRL) pre-treated with $scCO_2$ or ILs. The CRL activities were respectively 1.2- and 1.3-fold higher over the untreated ones after pre-treatment with sub- and $scCO_2$ under the pressure of 6 MPa and 15 MPa at 40 °C for 20 min.[21]

Other results that can be used for optimising enzyme catalyzed reactions in $scCO_2$ at the industrial level is the fact that IL molecules function as a coating layer and protect enzymes from denaturing conditions in supercritical CO_2. By analysis of root mean square deviation it was found that the enzyme has more native and stable conformation in $scCO_2$/IL system than in supercritical CO_2.[22] Stability of enzymes is an important and challenging subject in industrial applications of the biocatalysts. The effect of the $scCO_2$ treatment strongly depends on the experimental conditions, the nature and the source of the enzyme and, mainly, whether the enzyme is presented in a free or immobilized form where the characteristics of the support should be also taken into account.[23]

Although the enzyme stability and activity highly depends on enzyme species, its water content in the solution, and on the temperature and pressure of the reaction system, some enzymes, such as lipases, dehydrogenases, oxidases, phosphatases, amylases, peroxidases, esterases are well suited for reactions in $scCO_2$.[10,24–27] Structure of the enzyme might be altered during extreme conditions, causing their denaturation and loss of activity. However, with less adverse conditions, minor structural changes can be obtained, with altered enzyme activity and stability.

It was noticed that when some enzymes are treated with $scCO_2$ their activity is enhanced. Very good results were obtained for continuous reuse of immobilized α-amylase catalyzed hydrolysis of soluble starch where after repeated reuse of the enzyme when its activity was lower than the initial activity this was treated with $scCO_2$ to enhance activity of the enzyme. Four successful retreatment steps were achieved totalling up to seventeen reactions, where the activity of the beads could not be enhanced to a value higher than the initial activity of the untreated enzyme implying that additional enzymatic reactions can still be conducted as would be the case with the commercial enzyme.[28]

4.2.1 Effects of Temperature and Pressure

Fourier transfer infrared spectra, ultraviolet spectra, fluorescence spectra and scanning electron microscopy analyses demonstrate that some alterations in the secondary and tertiary structures, and surface morphology of the in $scCO_2$ treated enzymes may occur. This is the explanation why *e.g.* the activity and stability of pectinase were significantly improved under appropriate conditions in this medium. Significant increases in activity and stability of treated pectinase could be available with pressure lower than 15 MPa, whereas, temperature tends to reduce enzymatic activity and stability.[1]

With such results we come to a conclusion that parameters, such as pressure and temperature, at which an enzyme is treated in sub- or supercritical conditions play a very important role. Two important effects must be taken into account. First, enzyme activity changes with pressure and/or temperature change for different reasons and second, substrate solubility and diffusivity. Temperature and pressure can be used to influence density and transport properties, such as viscosity, thermal conductivity, diffusivity of supercritical fluids, which affects the solubility and transport of reactants and products to/from enzyme. Enzymes' structure is held together by hydrogen bonding, ionic and hydrophobic interactions, and by disulfide (S–S) bridges. Enzymes with disulfide bridges are in general more stable compared to enzymes without such covalent bonding. Change in $scCO_2$ pressure influences density-dependant properties, such as partition coefficient, dielectric constant and Hildebrandt solubility parameters. Those parameters regulate the activity, stability and specificity of enzymes.[29,30] Reaction parameters, such as rate constant can also be directly affected by pressure and temperature. Within the ranges of pressure (10 MPa to 40 MPa) and temperature (35 °C to 60 °C) that typically characterize the supercritical region of CO_2, an increase in pressure and/or a decrease in temperature may lead to a decrease in the enzyme turnover because the diffusion coefficients of the substrates migrating to the active sites of enzymes are affected. Temperature is one of the most important reaction parameter. It influences enzyme activity much more than pressure. Two effects are joined during an increase in reaction temperature: the reaction rate increases with higher temperature and enzyme activation/deactivation occurs. Within limits, activity improves with increasing temperature and ability to withstand a combination of temperature and pressure varies with enzymes.[5] As an example, in both atmospheric pressure and under supercritical conditions (30 MPa), the activity of Promod 144 (proteinase from *Carica papaya* latex) in $scCO_2$ was increased by an increase in temperature from 20 °C to 60 °C. The optimal temperature in $scCO_2$ was lower than at atmospheric pressure because of the co-extraction of water from the enzyme microenvironment by $scCO_2$.[29]

In $scCO_2$ optimal temperature is also connected with operational pressure. On pressure/temperature combination solubility of substrates and products are dependent. Pressure can modify the catalytic behaviour of enzyme by changing for example, the rate-limiting step or modulating the selectivity of the enzyme. If an enzyme is stable in a supercritical fluid its stability is usually not influenced by pressure for pressure ranges up to 30 MPa. However, reaction rates may be influenced by the pressure. In most cases a pressure increase acts positive for enzymatic reactions or there are no changes in the reaction rates. Pressure-induced deactivation of enzymes takes place mostly at pressure exceeding 150 MPa. Reversible pressure denaturation mostly occurs at pressures below 300 MPa and higher pressure needed to cause irreversible denaturation.

Reaction rate can be affected by changing the concentration of reactants and products, due to partitioning of reaction components between the two phases that are dependent on pressure. Therefore, pressure changes can adversely alter the enantioselectivity od enzymatic reactions.[31–33] Although there is no way to predict enzyme activity and enantioselectivity in $scCO_2$ this was successfully used in enantioselective enzymatic reactions, *e.g.* in kinetic resolution of b-hydroxychalcogenides promoted by CAL-B.[7]

4.2.2 pH of Medium and Formation of Carbonic Acid

Sensitivity of enzymes to pH changes depends on the type and source of the enzyme. Usually enzyme deactivation due to pH change is reversible. However, at drastical pH change irreversible deactivation of an enzyme may occur. For enzymes to function properly some amount of water must be present in the reaction mixture. Using $scCO_2$ as a reaction medium pH changes occur due to the formation of carbonic acid which may affect enzyme activity. Balaban *et al.* studied inactivation of pectinesterase in orange juice by $scCO_2$, where treatment with high pressure was accompanied by a lowering pH of medium that occurred due to the formation of carbonic acid from dissolution of carbon dioxide in water. The study suggests that the developed low pH of medium contributed to the low pectinesterase activity.[34] Later on, another study proves significant loss of pectinesterase activity due to lowering the pH of the medium (pH 2) with hydrochloric acid.[35] Nevertheless, some enzymes are active at quite low pHs.

4.2.3 Effect of Water Content

Water is crucial for enzymes and affects enzyme action in various ways: by influencing enzyme structure *via* non-covalent binding and disruption of hydrogen bonds, by facilitating reagent diffusion and by influencing the reaction equilibrium. Use of an enzyme in pure $scCO_2$ may lead to removal of the water, which is included or bonded to the enzyme. Therefore, the activity and stability of enzymes is highly dependent on the concentration of water. The quantity of the removed water is temperature and pressure depended.

It has already been reported that the effect of water on the activity of an immobilized lipase was reversible unless the concentration of water exceeded a specific value.[36] The effect of water on the activity of Lipozyme in $scCO_2$ was investigated. The study suggests that the water might lead to undesirable side reactions, such as hydrolysis of proteins or their conformational changes.[36–38] Another study reports of continuous acidolysis of triolein and stearic acid, carried out by immobilized lipase in $scCO_2$, where the moisture adsorption to the immobilized enzyme was influenced by the water concentration in $scCO_2$. The enzymatic reaction at increased pressure lead to a better partition of substrates to the $scCO_2$, which made the partition of the substrates to the immobilized lipase suppressed. The study presents optimal operating conditions of acidolysis with immobilized lipase in $scCO_2$, which are 50 °C, 16.9 MPa and an adsorbed-water concentration of 2 wt%.[39]

4.2.4 High-pressure Enzymatic Reactors

Model enzymatic reactions in $scCO_2$ were performed in different reactor types, not only batch-wise but also continuous-way where good results were obtained in high-pressure backed-bed reactors with the use of immobilized enzymes. Performing lipase-catalyzed production of *n*-octyl oleate by esterification of oleic acid with 1-octanol in dense CO_2 in the continuously operating bioreactor (Figure 4.3), a long-term enzyme lifetime was observed and a decrease of the Lipozyme activity was not registered over 50 days. Operating at the optimum reaction conditions, higher ester yield than those obtained in batch-mode was detected.[40] Similarly, the immobilized commercial lipase, Novozym 435, was very efficient in catalysing the synthesis of isoamyl acetate, a banana flavour ester, in $scCO_2$. It was active at pressures ranging from 8 MPa to 30 MPa, being stable not finding any loss of activity during one month of continuous operation.[41]

Synthesis of butyl acetate in an enzymatic packed bed reactor using lipase B, from *Candida antarctica*, immobilized by simple physical adsorption on porous pellets gave the productivity in $scCO_2$ (at 60 °C and under 12 MPa) of 501 mol min^{-1} g^{-1} pellets, which is about four times higher than what was

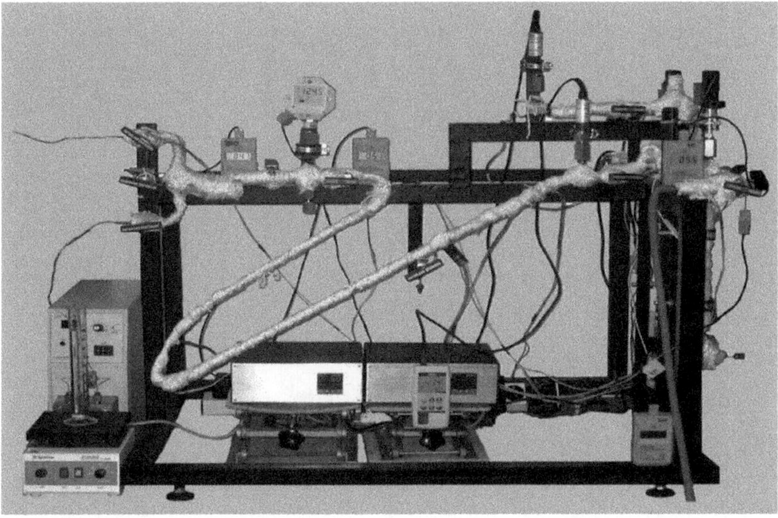

Figure 4.3 High-pressure continuous packed-bed reactor device.

obtained in hexane. Results showed that replacing hexane by $scCO_2$ in a continuous process increases ester productivity and reduces environmental impact, thus allowing making a more environmentally friendly process.[42]

Some studies were done with the enzymatic reactions performance in the high-pressure membrane reactors which enable the catalytic reaction, enzyme recovery and product isolation as a one-step process. The use of tubular membranes enables enzyme immobilization on their surface with the reduction of operational costs. Commercial expensive immobilized enzymes may be replaced by cheaper non-immobilized enzymes in this case. The application of tubular ceramic membranes in the high-pressure reaction system was studied on hydrolysis of carboxy-methyl cellulose (CMC) at atmospheric pressure and in biphasic $scCO_2$–H_2O medium and was catalyzed with covalent linked cellulase from *Humicola insolens* on the surface of ceramic membrane. A non-immobilized lipase-catalyzed hydrolysis of sunflower oil was performed in a high-pressure continuous enzymatic flat-shape membrane reactor in biphasic $scCO_2$–H_2O medium (Figure 4.4). Reactions carried out in biphasic medium gave higher productivity than reactions, performed at atmospheric pressure.[43] The advantage of high-pressure flat-shape membrane reactor is again in the possibility of using cheaper non-immobilized enzymes with longer life-time in the reactor.

4.3 Biomass Conversion in $scCO_2$

Biomass is regarded as an alternative energy source to fossil fuels, since in 2000 the energy derived from biomass corresponded to approximately 10% of the global energy supply.[44,45] Biomass as an energy source is considered

Figure 4.4 High-pressure continuous flat-shape membrane reactor.

sustainable, since it is CO_2 neutral in the life cycle, causing almost zero net emissions of CO_2. Furthermore, it contains negligible contents of nitrogen and sulphur. Among other various biomass utilization technologies, conversion of biomass into bio-oil by fast pyrolysis has been shown promising for engine fuels and high value chemicals.[46,47]

ScCO_2 may be involved in the biomass processing in three ways; as a pre-treatment agent, as a solvent for extraction or as a reaction medium. A study reports of separation of biomass pyrolysis oil by scCO_2 extraction, where high extraction pressure and relative low extraction temperature favoured in effectively reducing the water content in extracted bio-oil. The maximum extraction efficiency of scCO_2 extract bio-oil reached 88.6% on water-free basis at 30 MPa and 35 °C.[48]

Other study reports of guayule, a desert shrub harvested for commercial production of hypoallergenic latex, where the pre-treated biomass was subjected to enzyme hydrolysis. Supercritical method outperformed other methods and gave much higher overall sugar yields for guayule, resulting in 77% for glucose, 86% of for total reducing sugars. The scCO_2-based method was found to be very promising for pre-treatment of waste biomass of guayule.[49]

Cellulose the most abundant renewable resource on earth may be used for different purposes. One way is that cellulose and hemicellulose present in lignocellulose are converted to sugars by acid and/or enzymatic hydrolysis to be further fermented to produce ethanol or other valuable products. However, cellulose must be pre-treated to obtain sugar yield higher than 20%.

The $scCO_2$ pre-treatments of both hardwood and softwood with moisture contents of 40, 57, and 73% (w/w) together with enzymatic hydrolysis showed significantly higher final sugar yields compared to the thermal pre-treatments without $scCO_2$.[50] Similarly, for corn cob and corn stalk, the process of enzymatic hydrolysis was significantly improved by pre-treatment with $scCO_2$ and $scCO_2$/ultrasound combination. Compared with the untreated sample, the maximum total reducing sugar yield of corn cob using these two methods increased by 50% and 75%, respectively.[51] But the question is whether this procedure would be commercialized due to the high capital costs. Maybe the use of immobilized cellulase would be a path to the commercialization. It was shown that in silica aerogel matrix immobilized cellulase could be well reused for hydrolysis of carboxymethyl cellulose without any significant loss of activity for at least 15 reaction cycles at atmospheric pressure and at least 20 reaction cycles when high-pressure system CO_2–H_2O was used as a reaction medium.[52]

Another very important use of cellulose is the production of thermo-softening plastics from cellulose derivatives. Acyl esters of cellulose are used effectively in the production of fibers, plastics, films, membranes, modern coatings, cosmetics and drugs.[53,54] Due to highly crystalline structure of cellulose this must be pre-treated to permit chemical manipulation. Commercially, such products are prepared by the use of heterogeneous reaction systems containing tough and hazardous substances like pyridine-*N*,*N*-dimethylacetamide and acid chloride. To overcome this disadvantages a system at which fibrous cellulose was enzymatically acylated with vinyl laurate in $scCO_2$ with the formation of cellulose laurate was successfully designed.[55]

Like previous study, pre-treatment of lignocellulosic biomass was investigated as well, where rice hush was submitted to supercritical pre-treatment at 80 °C and 27 MPa with the aim to determine the effect on lignin content, crystallinity as well as enzymatic digestibility. The results show a lignin content reduction up to 90.6% for the sample with 75% moisture content using a water–ethanol mixture.[56]

When treating cellulosic materials with $scCO_2$ to increase the reactivity of cellulose by enhancing the rate and extent of cellulose hydrolysis, Zheng *et al.* show a significant increase of glucose yield when the temperature of the reaction reached supercritical 35 °C. Upon an explosive release of CO_2 pressure, the disruption of the cellulosic structure increases the accessible surface area of its substrate to enzymatic hydrolysis. A higher pressure allows supercritical fluid to penetrate deeper into cells with a more powerful explosion action to open up the pores of biomass. As mentioned before, the study of Zheng *et al.* shows that more glucose was produced from the cellulosic material after the $scCO_2$ pre-treatment compared to the material without the pre-treatment.[57]

Later, an enzymatic synthesis of soybean biodiesel using $scCO_2$ as solvent in continuous expanded-bed reactor was investigated, where commercial immobilized lipase Novozym 435 was used as a catalyst. Influence of the amount of enzyme, pressure, ethanol to soybean oil molar ratio and

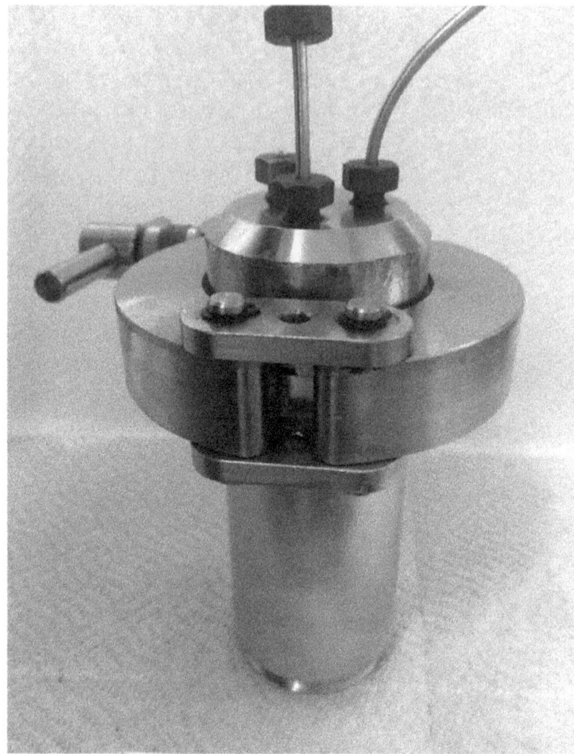

Figure 4.5 High-pressure bioreactor.

substrate to solvent mass ratio was studied. As reported, 94% conversion was reached after 60 min of reaction, where the process was conducted using 40 g of enzyme with reactor operation conditions at 20 MPa, 70 °C, soybean oil to ethanol molar ratio of 1 : 9 and substrate to solvent mass ration of 1 : 3. Thus showing a technically feasible continuous production of enzymatic biodiesel using $scCO_2$ as solvent.[58]

Another study using immobilized lipase Novozym 435 was reported, performing synthesis of fatty acid methyl esters (FAME) as biodiesel from corn oil. The reaction was performed in a batch $scCO_2$ bioreactor (Figure 4.5) at various enzyme loads (5% to 15%), temperatures (40 °C to 60 °C), substrate mole ratios (corn oil–methanol, 1 : 3–1 : 9), pressures (10 MPa to 30 MPa) and times (1 h to 8 h). The highest FAME content (81.3%) was obtained with the reaction performed at 15% enzyme load, 60 °C, 1 : 6 substrate mole ratio and 10 MPa in 4 h. Even though high yields are obtained in batch reactors, continuous reactors are found to be more efficient in terms of yield, ease of process and time. Nevertheless, batch $scCO_2$ bioreactors were found to be inhibited at lower levels of excess alcohol compared to continuous ones.[59]

4.3.1 Algal Biomass in $scCO_2$

Biodiesel, which is produced by transesterification of triglycerides has been considered as an alternative to petroleum diesel.[60,61] Using conventional feedstock, such as vegetable oils has negatively affected food security and increased food prices. In recent decades, microalgae as a nonedible feedstock has been considered due to its high amount of lipids content and high growth rate, as well as their capacity to grow in saline water and seasonal tolerance.[62–65] Their content of lipids is usually in range of 20–50% and can even be enhanced by varying growth conditions.[66,67] Also, microalgae with 30% lipids content have been reported to have ten times lipid productivity per hectare higher than the best oil producing crop, such as oil palm.[65,68] Microalgae have proved to have several functional properties, such as anticancer, antiviral, anti-oxidative, as well as anti-obesity activities[69–72] and are able to synthesize various molecules, such as for food, nutraceuticals and pharmaceutical applications.[66,73–76]

Various species of microalgae have already been cultivated for biodiesel production, giving a range of over 3000 different microalgal strains.[77] For microalgal biomass to be used for biodiesel production, extraction of lipids has to be performed. Over the past years many conventional lipid extraction techniques have been used, such as Bligh and Dyer,[78] Folch,[79] or Soxhlet[80] which are very time consuming, moreover the biomass processed is contaminated with used solvent, which limits additional biomass usage. Taking that into account, $scCO_2$ has showed promising alternative to organic solvents for lipids extraction. One of the most important advantages of $scCO_2$ is that after the extraction, the solvent can easily be separated from the extract by sample depressurizing, which allows no solvent residue remaining in the residual solid matrices. That advantage allows the leftover biomass to be further used in pharmaceutical and food applications (Figure 4.6).

Figure 4.6 Biomass from algae *Spirulina platensis.*

There are some studies that report of biodiesel production from microalgal biomass. An enzymatic production of biodiesel from microalgal lipids was investigated in a batch and integrated extraction-reaction systems, where the highest transesterification yield of 80% was obtained at 47 °C, 20 MPa, 35% enzyme loading and a 9 : 1 molar ration after 4 h reaction in a batch system.[81]

A wet extraction process of carotenoids and lipids from algae using liquefied (subcritical) dimethyl ether (DME) as solvent at around 0.59 MPa was investigated. For this liquefied DME extraction, a semi-continuous flow-type apparatus was used, into which raw material of algae (93.2% water content) was loaded. Extraction was performed at various pressures from 10 MPa to 40 MPa with a constant temperature of 60 °C, and at various temperatures from 40 °C to 70 °C with a constant pressure of 40 MPa. They report to obtain 58 $\mu g\ g^{-1}$ of dry weight algae at 40 MPa and 60 °C. Also, extraction yield increased at constant temperature of 60 °C, but with increasing pressure.[82]

Another study reports of lipids extraction from different algae being enzymatically transesterified to produce biodiesel in $scCO_2$. It was shown, that the conversion of biodiesel produced from microalgae lipids was 35% higher than that achieved using lamb fat in similar systems. The reaction was carried out at 50 °C, 20 MPa, 24 h reaction time with 30% enzyme loading (wt%) and a 4 : 1 methanol to lipid ratio.[83]

A study on the role of co-solvents in improving direct supercritical methanol transesterification of wet microalgal biomass under $scCO_2$ was performed. In this study, microalgal lipids were simultaneously extracted and converted to biodiesel under high pressure and temperature conditions without catalyst. Supercritical methanol was investigated, and the best performance was achieved by using methanol/wet biomass ratio of 8 : 1. Among the studied co-solvents, hexane was the most efficient in increasing crude extraction, with a hexane/biomass ratio of 6 : 1.[84]

4.4 Conclusion

The application of $scCO_2$ and SCFs in general has become a novel and advantageous technique for performing numerous enzyme-catalyzed reactions. Because of its gaseous nature of CO_2 under conventional conditions, $scCO_2$ has a big advantage over other SCFs. In that manner, any enzyme-catalyzed process performed in $scCO_2$ can profit in achieving only by changing the reaction conditions, such as changing pressure and changing temperature.

Application and incorporation of $scCO_2$ into performing industrially important enzymatic reactions and enzyme-based biomass catalyzed reactions is ever growing and becoming an important, green, and sustainable, more importantly even environmentally friendly medium.

Acknowledgements

The authors would like to thank Slovenian Research Agency and programme Separation Processes and Product Design (Contract No. P2-0046).

References

1. Y.-K. Peng, L.-L. Sun, W. Shi and J.-J. Long, *J. Cleaner Prod.*, 2016, **125**, 331–340.
2. S. Yan, B. Wang, Z. Wang, D. Hu, X. Xu, J. Wang and Y. Shi, *Biosens. Bioelectron.*, 2016, **80**, 34–38.
3. R.-H. Wang, H.-J. Huo, Z.-Y. Huang, H.-F. Song and H.-J. Ni, *J. Hydrodyn., Ser. B*, 2014, **26**, 226–233.
4. Y. Jiang, Y. Luo, Y. Lu, C. Qin and H. Liu, *Energy*, 2016, **97**, 173–181.
5. K. Rezaei, F. Temelli and E. Jenab, *Biotechnol. Adv.*, 2007, **25**, 272–280.
6. M. Habulin and Ž. Knez, *J. Chem. Technol. Biotechnol.*, 2001, **76**, 1260–1266.
7. C. G. Laudani, M. Habulin, Ž. Knez, G. Della Porta and E. Reverchon, *J. Supercrit. Fluids*, 2007, **41**, 92–101.
8. G. H. Prado, M. Khan, M. D. Saldaña and F. Temelli, *J. Supercrit. Fluids*, 2012, **66**, 198–206.
9. H. Shekarchizadeh, M. Kadivar, H. S. Ghaziaskar and M. Rezayat, *J. Supercrit. Fluids*, 2009, **49**, 209–215.
10. M. Habulin, S. Šabeder, M. A. Sampedro and Ž. Knez, *Biochem. Eng. J.*, 2008, **42**, 6–12.
11. M. Habulin, S. Šabeder and Ž. Knez, *J. Supercrit. Fluids*, 2008, **45**, 338–345.
12. A. Murakami, Y. Nakamura, K. Koshimizu, D. Takahashi, K. Matsumoto, K. Hagihara, H. Taniguchi, E. Nomura, A. Hosoda and T. Tsuno, *Cancer Lett.*, 2002, **180**, 121–129.
13. F.-H. Lin, J.-Y. Lin, R. D. Gupta, J. A. Tournas, J. A. Burch, M. A. Selim, N. A. Monteiro-Riviere, J. M. Grichnik, J. Zielinski and S. R. Pinnell, *J. Invest. Dermatol.*, 2005, **125**, 826–832.
14. D. L. Compton and J. W. King, *J. Am. Oil Chem. Soc.*, 2001, **78**, 43–47.
15. K. Sabally, S. Karboune, F. K. Yeboah and S. Kermasha, *Appl. Biochem. Biotechnol.*, 2005, **127**, 17–27.
16. K. Sabally, S. Karboune, R. St-Louis and S. Kermasha, *J. Am. Oil Chem. Soc.*, 2006, **83**, 101–107.
17. D. Ciftci and M. D. Saldaña, *J. Supercrit. Fluids*, 2012, **72**, 255–262.
18. M. Habulin, S. Šabeder, M. Paljevac, M. Primožič and Ž. Knez, *J. Supercrit. Fluids*, 2007, **43**, 199–203.
19. E. Garcia-Junceda, *Multi-step Enzyme Catalysis*, Wiley Online Library, 2008.
20. M. Primožič, S. Kavčič, Ž. Knez and M. Leitgeb, *J. Supercrit. Fluids*, 2016, **107**, 414–421.

21. Y. Liu, D. Chen and Y. Yan, *J. Mol. Catal. B: Enzym.*, 2013, **90**, 123–127.
22. H. Monhemi, M. R. Housaindokht, M. R. Bozorgmehr and M. S. S. Googheri, *J. Supercrit. Fluids*, 2012, **69**, 1–7.
23. R. Melgosa, M. T. Sanz, Á. G. Solaesa, S. L. Bucio and S. Beltrán, *J. Supercrit. Fluids*, 2015, **97**, 51–62.
24. X. Xia, Y.-H. Wang, B. Yang and X. Wang, *Biotechnol. Lett.*, 2009, **31**, 83–87.
25. K. Stránský, M. Zarevúcka, Z. Kejík, Z. Wimmer, M. Macková and K. Demnerová, *Biochem. Eng. J.*, 2007, **34**, 209–216.
26. M. Leitgeb, M. Čolnik, M. Primožič, P. Zalar, N. G. Cimerman and Ž. Knez, *J. Supercrit. Fluids*, 2013, **78**, 143–148.
27. Ž. Knez, S. Kavčič, L. Gubicza, K. Bélafi-Bakó, G. Németh, M. Primožič and M. Habulin, *J. Supercrit. Fluids*, 2012, **66**, 192–197.
28. D. Senyay-Oncel and O. Yesil-Celiktas, *J. Mol. Catal. B: Enzym.*, 2013, **91**, 72–76.
29. M. Habulin, M. Primožič and Ž. Knez, *J. Supercrit. Fluids*, 2005, **33**, 27–34.
30. G. Nagesha, B. Manohar and K. U. Sankar, *J. Supercrit. Fluids*, 2004, **32**, 137–145.
31. T. Matsuda, T. Harada, K. Nakamura and T. Ikariya, *Tetrahedron: Asymmetry*, 2005, **16**, 909–915.
32. J. C. Erickson, P. Schyns and C. L. Cooney, *AIChE J.*, 1990, **36**, 299–301.
33. T. Matsuda, K. Watanabe, T. Harada and K. Nakamura, *Catal. Today*, 2004, **96**, 103–111.
34. M. O. Balaban, A. G. Arreola, M. Marshall, A. Peplow, C. I. Wei and J. Cornell, *J. Food Sci.*, 1991, **56**, 743–746.
35. J. Owusu-Yaw, M. Marshall, J. Koburger and C. Wei, *J. Food Sci.*, 1988, **53**, 504–507.
36. T. Dumont, D. Barth and M. Perrut, Continuous synthesis of ethyl myristate by enzymatic reaction in supercritical carbon dioxide, in 2nd International Symposium on Supercritical Fluids Boston, 1991.
37. A. Marty, W. Chulalaksananukul, R. Willemot and J. Condoret, *Biotechnol. Bioeng.*, 1992, **39**, 273–280.
38. Y. M. Chi, K. Nakamura and T. Yano, *Agric. Biol. Chem.*, 1988, **52**, 1541–1550.
39. M. Hakoda, N. Shiragami, A. Enomoto and K. Nakamura, *Bioprocess Biosyst. Eng.*, 2002, **24**, 355–361.
40. C. G. Laudani, M. Habulin, Ž. Knez, G. Della Porta and E. Reverchon, *J. Supercrit. Fluids*, 2007, **41**, 74–81.
41. M. Romero, L. Calvo, C. Alba, M. Habulin, M. Primožič and Ž. Knez, *J. Supercrit. Fluids*, 2005, **33**, 77–84.
42. J. Escandell, D. Wurm, M. Belleville, J. Sanchez, M. Harasek and D. Paolucci-Jeanjean, *Catal. Today*, 2015, **255**, 3–9.
43. M. Primožič, M. Paljevac and Ž. Knez, *Desalination*, 2009, **241**, 14–21.
44. J. Jones and K. Semrau, *Biomass*, 1984, **5**, 109–135.
45. Y. Sun and J. Cheng, *Bioresource Technol.*, 2002, **83**, 1–11.

46. A. Bridgwater, D. Meier and D. Radlein, *Org. Geochem.*, 1999, **30**, 1479–1493.
47. D. Chiaramonti, A. Oasmaa and Y. Solantausta, *Renewable Sustainable Energy Rev.*, 2007, **11**, 1056–1086.
48. W. Jinghua, C. Hongyou, W. Shuqin, Z. Shuping, W. Lihong, L. Zhihe and Y. Weiming, *Smart Grid Renewable Energy*, 2010, **1**, 98–107.
49. N. Srinivasan and L.-K. Ju, *Bioresour. Technol.*, 2010, **101**, 9785–9791.
50. K. H. Kim and J. Hong, *Bioresour. Technol.*, 2001, **77**, 139–144.
51. J. Z. Yin, L. D. Hao, W. Yu, E. J. Wang, M. J. Zhao, Q. Q. Xu and Y. F. Liu, *Chin. J. Catal.*, 2014, **35**, 763–769.
52. M. Paljevac, M. Primozic, M. Habulin, Z. Novak and Z. Knez, *J. Supercrit. Fluids*, 2007, **43**, 74–80.
53. K. J. Edgar, C. M. Buchanan, J. S. Debenham, P. A. Rundquist, B. D. Seiler, M. C. Shelton and D. Tindall, *Prog. Polym. Sci.*, 2001, **26**, 1605–1688.
54. K. J. Edgar, *Cellulose*, 2007, **14**, 49–64.
55. S. Gremos, *Bioresour. Technol.*, 2012, **115**, 96–101.
56. L. D. Serna, C. O. Alzate and C. C. Alzate, *Bioresour. Technol.*, 2016, **199**, 113–120.
57. Y. Zheng, H. Lin and G. T. Tsao, *Biotechnol. Prog.*, 1998, **14**, 890–896.
58. T. S. Colombo, M. A. Mazutti, M. Di Luccio, D. de Oliveira and J. V. Oliveira, *J. Supercrit. Fluids*, 2015, **97**, 16–21.
59. O. N. Ciftci and F. Temelli, *J. Supercrit. Fluids*, 2013, **75**, 172–180.
60. G. Knothe, *Prog. Energy Combust Sci.*, 2010, **36**, 364–373.
61. A. Demirbaş, *Energ. Convers. Manage.*, 2003, **44**, 2093–2109.
62. K. Vijayaraghavan and K. Hemanathan, *Energ. Fuel.*, 2009, **23**, 5448–5453.
63. V. Patil, K.-Q. Tran and H. R. Giselrød, *Int. J. Mol. Sci.*, 2008, **9**, 1188–1195.
64. X. Meng, J. Yang, X. Xu, L. Zhang, Q. Nie and M. Xian, *Renewable Energy*, 2009, **34**, 1–5.
65. Y. Chisti, *Biotechnol. Adv.*, 2007, **25**, 294–306.
66. P. Spolaore, C. Joannis-Cassan, E. Duran and A. Isambert, *J. Biosci. Bioeng.*, 2006, **101**, 87–96.
67. M. Adamczak, U. T. Bornscheuer and W. Bednarski, *Eur. J. Lipid Sci. Technol.*, 2009, **111**, 800–813.
68. A. Demirbas, *Prog. Energy Combust. Sci.*, 2007, **33**, 1–18.
69. I. Wijesekara, R. Pangestuti and S.-K. Kim, *Carbohydr. Polym.*, 2011, **84**, 14–21.
70. S.-K. Kim and I. Wijesekara, *J. Funct. Foods*, 2010, **2**, 1–9.
71. S. A. Pomponi, *J. Biotechnol.*, 1999, **70**, 5–13.
72. D.-S. Choi, Y. Athukorala, Y.-J. Jeon, M. Senevirathne, K.-R. Cho and S.-H. Kim, *Prev. Nutr. Food Sci.*, 2007, **12**, 65–73.
73. M. Plaza, A. Cifuentes and E. Ibáñez, *Trends Food Sci. Technol.*, 2008, **19**, 31–39.

74. R. L. Mendes, B. P. Nobre, M. T. Cardoso, A. P. Pereira and A. F. Palavra, *Inorg. Chim. Acta*, 2003, **356**, 328–334.
75. C. Crampon, O. Boutin and E. Badens, *Ind. Eng. Chem. Res.*, 2011, **50**, 8941–8953.
76. O. Pulz and W. Gross, *Appl. Microbiol. Biotechnol.*, 2004, **65**, 635–648.
77. J. Sheehan, T. Dunahay, J. Benemann and P. Roessler, *A look back at the US Department of Energy's aquatic species program: biodiesel from algae*, National Renewable Energy Laboratory, 1998.
78. E. G. Bligh and W. J. Dyer, *Can. J. Biochem. Physiol.*, 1959, **37**, 911–917.
79. J. Folch, M. Lees and G. Sloane-Stanley, *J. Biol. Chem.*, 1957, **226**, 497–509.
80. E. Ryckebosch, K. Muylaert and I. Foubert, *J. Am. Oil Chem. Soc.*, 2012, **89**, 189–198.
81. H. Taher, S. Al-Zuhair, A. H. Al-Marzouqi, Y. Haik and M. Farid, *Biochem. Eng. J.*, 2014, **90**, 103–113.
82. M. Goto, H. Kanda and S. Machmudah, *J. Supercrit. Fluids*, 2015, **96**, 245–251.
83. H. Taher, S. Al-Zuhair, A. Al-Marzouqi, Y. Haik and M. Farid, *Renewable Enery.*, 2015, **82**, 61–70.
84. H. A. Najafabadi, M. Vossoughi and G. Pazuki, *Bioresour. Technol.*, 2015, **193**, 90–96.

CHAPTER 5

Direct Hydrolysis of Biomass Polymers using High-pressure CO_2 and CO_2–H_2O Mixtures

ANA RITA C. MORAIS[a,b] AND RAFAL M. LUKASIK*[b]

[a] Unidade de Bioenergia, Laboratório Nacional de Energia e Geologia, I.P., Estrada do Paço do Lumiar 22, 1649-038 Lisboa, Portugal;
[b] LAQV-REQUIMTE, Departamento de Química, Faculdade de Ciências e Tecnologia, Universidade NOVA de Lisboa, Caparica, Portugal
*Email: rafal.lukasik@lneg.pt

5.1 Introduction

Throughout the world, non-renewable fossil resources (petroleum, coal and natural gas) have been used to produce a wide range of fuels, chemicals and materials. However, the massive consumption of petroleum-based feedstocks has led to their limited availability and together with frightening environmental problems have increased the need for new feedstocks.[1] Increasing concerns regarding these issues have resulted in an ever-increasing shift of global energy strategies to develop new sources of "green" products and alternative chemical technologies as well.[2] According to the Renewable Energy Sources and Climate Change Mitigation (SRREN), 80% of the worldwide energy supply will be obtained from renewable sources within the next four decades.[3,4] Currently, biomass has been acknowledged as the most sustainable carbon-based feedstock for manufacturing of liquid fuels and commodity chemicals, by developing more sustainable technologies.[5–7] Biomass is a non-fossil, renewable, abundant and carbon neutral

Green Chemistry Series No. 48
High Pressure Technologies in Biomass Conversion
Edited by Rafał M. Łukasik

Published by the Royal Society of Chemistry, www.rsc.org

feedstock. Besides all biomass benefits, it is well-known that the use of 1st generation of biomass (*e.g.*, wheat, rice, corn, potato, *etc.*) is not the best choice to produce fuels and chemicals since it interferes, either directly or indirectly, with food and feed end-uses.[8] In this context, an increasing focus has been placed on the use of 2nd generation of biomass feedstocks (*e.g.*, lignocellulose) to replace the current non-renewable fossil-based products. At this point, lignocellulosic biomass, which is the most abundant, available locally and accessible in most of world regions, and woefully underutilized source of renewable feedstock worldwide, plays a critical role.[9] Several literature studies have demonstrated that lignocellulosic biomass has a vast potential to be sustainably converted, through a number of different processes, into a mixture of valuable products including biofuels, chemicals and materials.[9,10] In this matter, lignocellulosic biomasses have numerous benefits over other biomass feedstocks because they are non-edible parts of plants and thus, do not interfere with food production.[11] Furthermore, agro-industrial and forestry residues instead of being used as raw materials for the manufacture of a number of valuable products[12] are left in large quantities on the field or landfilled every year causing severe environmental and economic problems. Additionally, lignocellulosic biomass has been considered as an abundant carbon-neutral renewable source which can reduce CO_2 emissions and environment pollution.[13] Lignocellulosic biomass is made up of polymer forms, cellulose, hemicelluloses and lignin that can be used as substitute of petroleum-based polymers due to its eco-friendly features, such as bio-degradability and renewability. However, the development of effective technologies for the conversion of lignocellulosic biomass into added-value products remains a remarkable challenge, essentially due to its inherent complex and recalcitrant structure. High cellulose crystallinity, non-permeability of lignin and encapsulation of cellulose by the hemicellulose-lignin matrix give robustness to lignocellulosic biomass.[14] This structural characteristic makes lignocellulosic biomass very tough to be chemically converted.[15] Hence, to change the physical and chemical properties of lignocellulosic biomass, the employment of hydrolysis technologies is indispensable, which is very costly and energy-demanding.[16] Although the majority of lignocellulosic biomass is available in high amounts and often has low or even null economic value, the true challenge is to convert lignocellulosic biomass into valuable products with high yield and selectivity and in a sustainable manner.

In this context, high-pressure CO_2 and CO_2–H_2O mixture are interesting alternatives and play an important role in diverse steps of biomass processing technologies. The hydrolysis of biomass polymers, carried out in presence of CO_2 and water, which can be even from biomass moisture, has been considered as a serious alternative to conventional acid-based technologies. This chapter focuses on direct hydrolysis of a wide-range of polymers present in lignocellulosic biomass using high-pressure CO_2 and CO_2–H_2O mixtures, into their respective components that can be used as pivot-compounds to replace petroleum-based products.

5.1.1 Lignocellulosic Biomass Polymers

The worldwide production of lignocellulosic biomass is estimated of 10^{10} million tons.[17] Generally, lignocellulosic biomass consists mainly of cellulose, hemicelluloses and lignin and small amounts of secondary components, such as proteins, waxes, terpenoids, phenolic substituents, extractives, such as waxes and minerals.[18,19] Table 5.1 summarizes the chemical composition of selected examples of lignocellulosic biomasses.

Additionally, considering the remarkable amounts of starch frequently found in biomass materials (wastes from starch processing industries, *e.g.* sago pith waste), starchy biomasses are important sources of upgradable biopolymers.[20] Depending on specie of biomass, growth stage, age and environmental conditions, these components are organized into complex non-uniform three-dimensional structures and present in various proportions.

The main polymer found in lignocellulosic biomass is cellulose and it is present in an organized fibrous structure. The linear polymer consists in an intra/intermolecular hydrogen bonding networks linking firmly the monomers of glucose as shown in Scheme 5.1. In wood-derived biomass, cellulose wires are entwined with rigid and water impermeable matrices of lignin.

Table 5.1 Chemical composition (% dry weight) of selected lignocellulosic biomasses.

Lignocellulosic biomass		Cellulose (%)	Hemicelluloses (%)	Total lignin (%)	Ref.
Hardwood	Oak	45	25	24[a]	95
	Poplar	45	23	20	96
	Eucalyptus	54	18	22	96
Softwood	Spruce	46	23	28	96
	Pine	43	22	28	96
Agricultural	Wheat straw	39	25	18	45
	Rice straw	31	22	13	97
	Sugarcane bagasse	43	31	11	98
	Corn cob	34	32	6	99
	Corn stover	38	26	17	100
Grasses	Switchgrass	40	25	26[a]	101

[a]Sum of acid insoluble and soluble lignin.

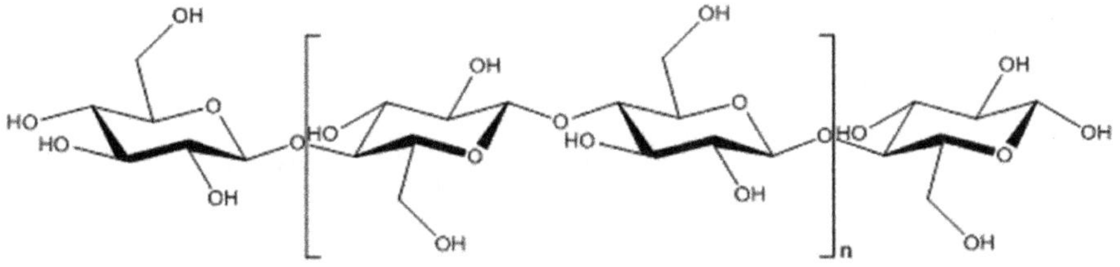

Scheme 5.1 The chemical structure of cellulose.

This network arrangement leads to a rigid and crystalline structure of cellulose making it very difficult to be converted into its monomer components under mild or enzymatic conditions.[21]

Considering that almost 50% of the organic carbon in the biosphere is present in form of cellulose, in the context of lignocellulosic biomass valorization, cellulose plays an important role and because of this a special attention is given to this polymer.[9,13]

Hemicelluloses are typically the second most abundant component of lignocellulosic biomass. These polymers are located in the primary and secondary plant cell walls to form a complex bond network that provide structural rigidity by linking cellulose fibres into microfibrils and cross-linking with lignin.[13,22] Unlike cellulose, hemicelluloses (Scheme 5.2) display both branched and amorphous structure, which is composed of various heteropolymers, including xylan, xyloglucan, glucuronoxylan, arabinoxylan and glucomannan.[23–25] The chemical composition of heteropolymers depends on the source of hemicelluloses; hardwood and agricultural hemicelluloses are composed mainly of xylans, whereas softwood hemicelluloses contain mostly glucomannans.[23] The rate of dissolution depends on the sugar type and decreases in the following order: mannose, xylose, glucose, arabinose and galactose. The monosaccharide dissolution is also highly dependent on reaction temperature and under hydrothermal conditions the dissolution starts at 180 °C.[26] It is important to highlight that besides the process temperature the dissolution of hemicellulose compounds is dependent on other operational parameters, such as moisture content and pH value.[23] Therefore, the optimal operation parameters for the conversion of hemicellulose fraction to its component monomers should be a compromise between the reaction condition severe enough to achieve good reaction selectivities and yields, and at the same time, as mild as possible to be economically and energetically favourable.

Lignin is the third most abundant polymer in nature present in the cellular wall. It is a three-dimensional polymer composed by phenylpropanoid units (*p*-coumaryl, coniferyl and sinapyl alcohols) as shown in Scheme 5.3.[23] The corresponding phenylpropanoid units in the lignin are known as *p*-hydroxyphenyl (H), guaiacyl (G) and syringyl (S) units, respectively.[27,28]

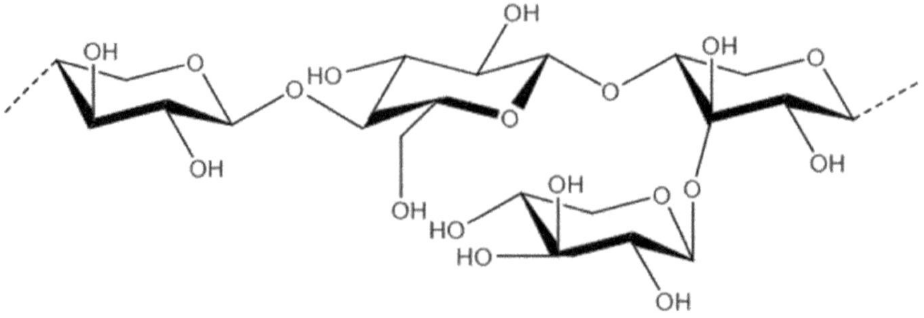

Scheme 5.2 The chemical structure of hemicellulose.

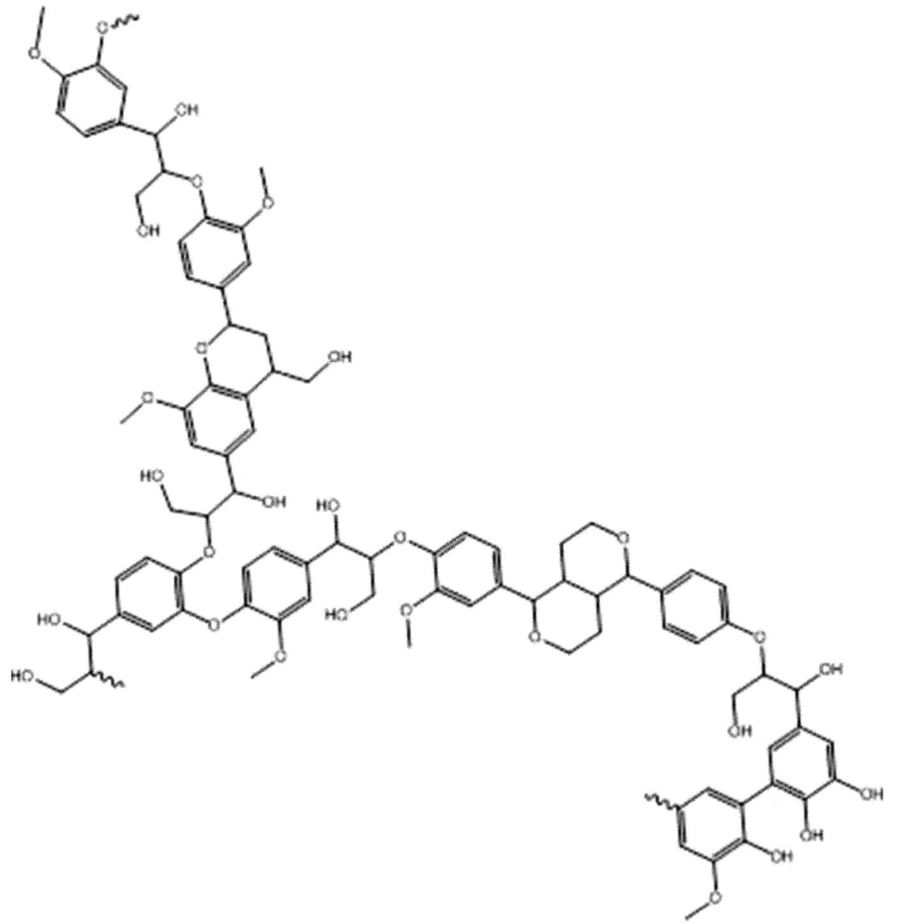

Scheme 5.3 Representative fragment of lignin illustrating its remarkable complexity.

The abundance of these units in lignin depends on plant species and tissue.[27,28] Because of lignin association with microfibrils of cellulose, it acts as cellular "glue" providing rigidity, non-permeability and resistance to cell wall against microbial attack and oxidative stress.[29] Lignin conversion has a great potential because its unique structure and chemical properties permit to produce a wide-range of valuable chemicals, mostly aromatic compounds. Thus, lignin has been considered the most important resource of aromatics in the context of the bio-based economy. Because of this, numerous technologies and strategies have been reported for the conversion of lignin towards valuable chemicals and fuels.[30]

Starch is an α-linked carbohydrate consisting of two glucose-based polymers present in various contents depending on the plant source: 10–20% 1,4-α-linked linear amylose and 80–90% both 1,4-α-linked linear and

1,6-α-linked branched amylopectin.[31] Due to its reduced crystallinity (crystallinity index of 35%), starch is easier to be hydrolyzed than cellulose even under mild hydrothermal conditions.

Proteins are one of the minor components present in lignocellulosic biomass, comprising no more than 1% (dry wt.) of woody biomass.[32] Although the general low content of proteins in lignocellulosic biomass, the building blocks of proteins—amino acids, have higher economic value than other biomass constituents. Amino acids play an important role in various industrial applications, such as food (as additives), medical,[33] cosmetic sectors.[34] Additionally, proteins can also be used as raw materials for the production nitrogen-containing materials due to the presence of amine group ($-NH_2$).[35] However, the conversion of biomass-derived proteins into amino acids has rarely been investigated. The simplest way to produce platform chemicals from proteins is *via* hydrolysis of peptides bonds, which are C–N linkage between the carboxyl and amine groups, producing a mixture of various amino acids. It is well-known that C–N linkages is highly susceptible to hydrolysis and degradation in hydrothermal processes, leading to amino acids yields significantly below those obtained in conventional acid-based processes.[36]

5.2 High-pressure CO_2 and CO_2–H_2O Mixture in the Hydrolysis of Biomass

5.2.1 Fundamentals

For over 40 years, high-pressure fluids (*i.e.* sub/supercritical fluids) have received an increasing attention as interesting alternatives to common solvents leading to the opening of a wide range of applications. Taking into consideration the green chemistry principles, the most promising high pressure chemicals are carbon dioxide and water, which are considered as renewable and non-flammable solvents. Additionally, for reactions performed under or near supercritical conditions, water has intriguing chemical properties and carbon dioxide can act as useful solvent in liquid phase. Research on the importance of water in hydrolysis of lignocellulosic biomass shows that water is essential, once it increases the final sugar yields, improving the efficiency of the process.[37] Under high pressures (200 bar or above) and temperatures between 160 and 250 °C, the gas phase is mainly composed of CO_2 (up to 30 mol% of water) and has a density comparable to a liquid, whereas liquid (reactional) phase consists mostly of water (up to 2 mol% of CO_2).[37] Under these conditions, carbohydrates undergo a hydrolysis process in a reactional phase composed by liquid hot water and dissolved CO_2 where the latter acts as acid catalyst promotor.[36] Additionally, high-pressure CO_2 leads to high diffusivities and the swelling effect on plant biomass.[38] CO_2 dissolved in water, in molecular form, leads to the *in situ* formation of carbonic acid (H_2CO_3), which dissociates in two stages into the bicarbonate ion (HCO_3^-), and later, into the carbonate ion (CO_3^{2-}), and the respective hydronium ions (H_3O^+) according to the following equation:

$$CO_2 + 2H_2O \leftrightarrow HCO_3^- + H_3O^+;\ H_2O \leftrightarrow CO_3^{2-} + H_3O^+.$$

Thus, the dissolution of CO_2 in water is believed to promote the acid-catalyzed hydrolysis of biomass polymers (like dilute-acid hydrolysis mechanism) by carbonic acid formation. The formation of carbonic acid increases the hydronium ion concentration in the medium, decreasing the pH value of reactional system, for instance between 2.8 and 3.0 under pressures of 70–200 bar and temperatures of 25–70 °C.[39] King and Srinivas plotted the reported literature results[40–43] related to the CO_2 solubility in water at different temperatures and CO_2 pressures as it is depicted in Figure 5.1.[44]

It is important to highlight that acidity of the reactional medium is highly dependent on CO_2 pressure and temperature because the solubility of CO_2 in water depends on both reaction parameters. Besides the aforementioned effect of carbonic acid, it also causes the hydrolysis of acetyl groups into acetic acid. Under CO_2 presence, the hydrolysis of acetyl groups is enhanced when compared to those obtained, for example, in liquid hot water. This way, CO_2 contributes to the formation of more acidic reactional environment and adds value to the hydrolytic process in two distinct ways: firstly, enhancement of hemicellulose polymer hydrolysis into

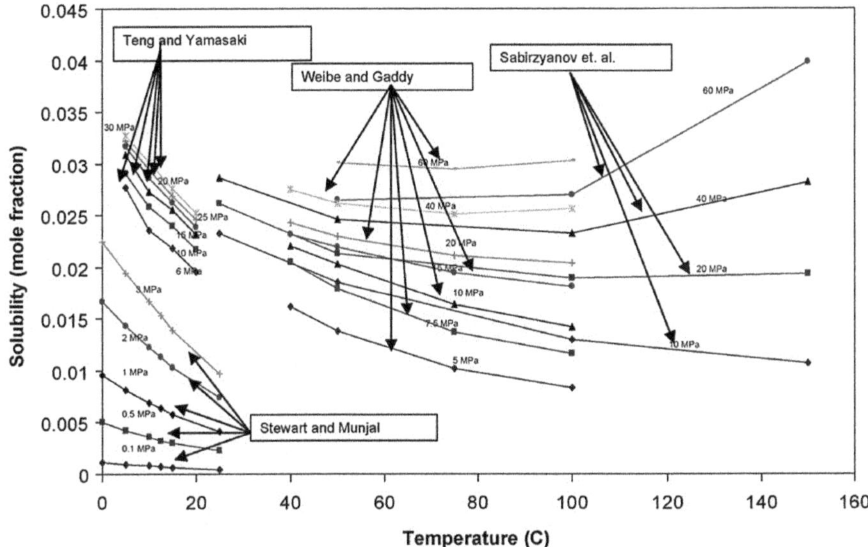

Figure 5.1 Solubility of CO_2 (mole fraction) as a function of temperature and CO_2 pressure.
Reprinted from J. W. King and K. Srinivas, Multiple unit processing using sub- and supercritical fluids, *J. Supercrit. Fluids*, **47**, 598–610,[44] Copyright (2009), with permission from Elsevier.

corresponding sugars,[45] and secondly, increasing the enzymatic digestibility of cellulose polymer.[46] It is clear that high-pressure CO_2–H_2O mixtures offer similar benefits to those found for either liquid hot water or acid-catalyzed methodologies, without the drawbacks typical for mineral acid catalysts, such as the need of neutralization and separation processes. Furthermore, the acidity of reactional medium does not constitute an environmental problem because CO_2 is practically immiscible in water under atmospheric conditions allowing to increase the pH value of the solution and recovery of CO_2.

5.2.1.1 *The Reaction Severity with High-pressure CO_2–H_2O Mixtures*

The severity factors provide an easy way to evaluate and to compare the results obtained under various reaction conditions (*i.e.* temperature, CO_2 pressure and reaction time). To assess the influence of CO_2 on the process severity, where the determination of pH value is crucial, a combined severity was proposed[47] as following: $CS_{P_{CO_2}} = \log(R_0) - \text{pH}$, where R_0 is the severity factor.[48] The solubility of CO_2 in water and, thus the dissociation of H_2CO_3 affects the concentration of hydronium ions in the mixture. The solubility of CO_2 in water can be described using Henry's constant as long as temperature is far from supercritical condition. Similarly to solubility of CO_2 in water, dissociation of H_2CO_3 in water is also depend on reaction temperature and its constant can be described by the following question[47] $\text{p}K_{a1} = \dfrac{2382.3}{T} - 8.513 + 0.02194T$, where $\text{p}K_{a1}$ is the negative decimal logarithm value of first dissociation constant of acid (K_{a1}) and T is temperature (K). Considering the effect of reaction temperature and partial pressure of CO_2, van Walsum[47] proposed the following equation that estimates the pH value of CO_2–H_2O mixture $\text{pH} = (8.00 \times 10^{-6})T^2 + 0.00209 \times T - 0.216 \times \ln(p_{CO_2}) + 3.92$, where the temperature is from 100 to 250 °C and the partial pressure of CO_2 is up to 151.9 bar. To evaluate the combined severity factor at specific reaction temperature, time and partial pressure of CO_2, the combined severity factor ($CS_{P_{CO_2}}$) is as following:[47] $CS_{P_{CO_2}} = \log(R_0) - ((8.00 \times 10^{-6})T^2 + 0.00209 \times T - 0.216 \times \ln(p_{CO_2}) + 3.92)$, where $CS_{P_{CO_2}}$ is the combined severity factor of the reaction in presence of CO_2, T is temperature (°C) and p_{CO_2} is partial pressure of CO_2 (atmospheres).

The proposed combined severity factor integrates all the most important reaction parameters affecting the final sugar production yield in the high-pressure CO_2–H_2O mixture. It is important to underline that the combined severity factor aims to compare the results obtained under different process conditions. Nevertheless, very often, lack of complete information regarding the pH value, process parameters such as heating profile and number of CO_2 moles (conditions under which CO_2 was charged) into the reactor make the comparison of literature results a challenging task.

5.3 Hydrolysis of Biomass-derived Polymers

The hydrolysis of biomass carbohydrates can be performed in acid[49] or enzymatic-catalyzed[50] processes. The disadvantages emerging from the addition of mineral acids (corrosion problems and the need of acid neutralization), or enzymes (long hydrolysis times), can be avoided by the substitution of them by hydrothermal processes. High-temperature liquid water as reaction medium has been considered mainly because water is naturally present in biomass and is one of the most eco-friendly solvent contrary to those used in nowadays practice. Because of this, hydrothermal technologies with or without addition of catalysts have been extensively investigated in biomass processing, for the production of a wide range of valuable products, such biofuels, chemicals[51] and biomaterials.[13]

The disadvantages emerging from the use of catalysts including mineral acids and heterogeneous catalysts, can be avoided by the substitution of acidic compounds with CO_2. Since CO_2 dissolved in water increases the availability of hydronium ions, the CO_2-catalyzed hydrolysis of biomass polymers has been investigated as alternative to homo- (mineral acids) or heterogeneous catalysts.[45–47,52–55]

5.3.1 Cellulose

Despite the huge potential of high-pressure CO_2–H_2O technology in the hydrolysis of cellulose, only a few works have been published to date. In the presence of water at high temperature and pressure, cellulose undergoes hydrolysis into C_6-sugars oligomers (*e.g.* cellohexaose, cellopentaose, cellotetraose, cellotriose, and cellobiose) and monomers (glucose and fructose). Due to the severe reaction conditions required for the hydrolysis of cellulose, glucose itself can be subject of undesired advanced hydrolysis into various degradation products, such as pyruvaldehyde, glyceraldehyde and 5-hydroxymethylfurfural (5-HMF).[56] As suggested by Sasaki *et al.* the products of cellulose hydrolysis are highly dependent on the reaction severity.[57] The results of their work demonstrate that cellulose was quickly converted into oligosaccharides, monomers and other low-molecular-weight products at temperature and pressure above critical point of water. Additionally, the same authors suggested that cellulose hydrolysis rate in subcritical water is slower than in supercritical water, yielding high glucose productivity.[57] The influence of acidification of CO_2 on both cellulose hydrolysis kinetics and product formation was studied by Brunner and co-workers and compared to the hydrolysis performance in pure water over a temperature range of 240–280 °C.[55] They found that the addition of CO_2 had a positive effect on cellulose liquefaction at 240 °C, however for temperatures above 260 °C, no further significant enhancement was obtained, as depicted in Figure 5.2. This less pronounced catalytic effect of CO_2 with an increase of temperature can be related to the associated decrease of pH of the CO_2-saturated water medium occurring as temperature increases. This leads to lower rate

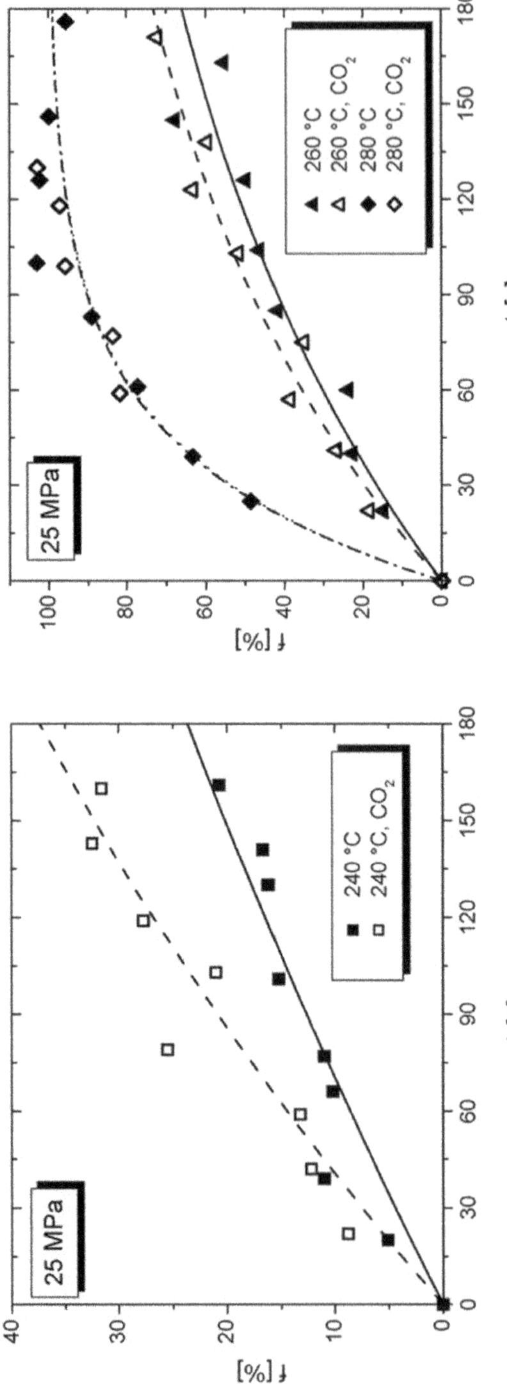

Figure 5.2 Degree of liquefaction of cellulose in pure and 100% CO_2-saturated water at 240, 260 and 280 °C. Reprinted from T. Rogalinski, K. Liu, T. Albrecht and G. Brunner, Hydrolysis kinetics of biopolymers in subcritical water, *J. Supercrit. Fluids*, **46**, 335–341,[55] Copyright (2008) with permission from Elsevier.

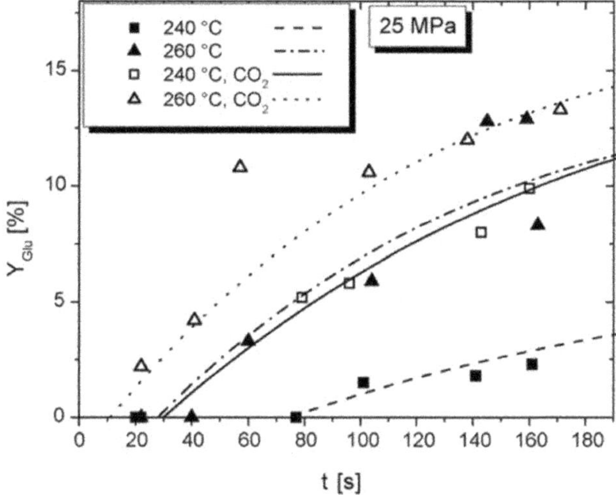

Figure 5.3 Influence of dissolved CO_2 in water on glucose yields from cellulose at different temperatures and reaction times.
Reprinted from T. Rogalinski, K. Liu, T. Albrecht and G. Brunner, Hydrolysis kinetics of biopolymers in subcritical water, *J. Supercrit. Fluids*, **46**, 335–341,[55] Copyright (2008) with permission from Elsevier.

improvement than those obtained in reactions carried out at higher temperatures.[55] Moreover, the effect of CO_2 addition on the product formation was also studied by the same group.[55] A superior production of glucose monomers from cellulose during CO_2-saturated water in the same aforementioned reaction conditons range was reported and it is shown in Figure 5.3. It is notorious that the onset glucose formation was shifted towards shorter reaction times due to faster cleavage of β-1,4-glycoside bonds facilitated by the *in situ* formation of carbonic acid. At these conditions glucose was formed at higher yields than those obtained for pure water but this improvement was less pronounced for temperature $\geq$260 °C.[55] Furthermore, the same authors indicated that solubility of CO_2 at set reaction temperature and reaction time plays an important role in cellulose hydrolysis, since undissolved CO_2 does not have any effect on the reaction progress. This means that CO_2 has a great importance on the acidification of the medium, but for a set CO_2 pressure, the pH of reactional medium is highly dependent on the reaction conditions applied.

5.3.2 Hemicelluloses

Hemicelluloses is a mixture of polysaccharides and due to lack of repeating β-1,4-glycosidic bonds do not present crystalline structure,[23] making it more susceptive to hydrothermal processing than crystalline cellulose. For instance, Liu and Wyman reported that more than 90% of original xylan present in corn stover was hydrolyzed under hydrothermal batch conditions at 220 °C for 16 min.[58] The number of studies on hydrolysis of hemicellulose in presence of CO_2 are increasing, being evident the growing interest of this technology in the production of C_5-sugars from biomass (Table 5.2). The addition of CO_2, as catalyst in hydrothermal technologies, represents an asset since it leads to use lower temperatures and shorter reaction times in comparison to CO_2-free water processes. van Walsum demonstrated that the *in situ* formation of carbonic acid has a hydrolytic effect on bench wood-derived xylan, promoting a slight enhancement in formation of xylose monomers and oligomers, with low degree of polymerization, when compared to those with liquid hot water process.[47] The same group also suggested that the CO_2 presence hinders the formation of organic acids.[52] The production of organic acids is highly dependent on the severity of the experimental conditions whereby, it is expected that high-pressure CO_2–H_2O, like other hydrolytic technologies, is also able to produce these compounds at certain reaction conditions. It is important to stress out that the beneficial chemical effects of high-pressure CO_2 only occur in the presence of water in the system, which might come from the biomass moisture. The benefits of high-pressure CO_2–H_2O technology instated of water-only reaction were also reported by Miyazawa and Funazukuri.[59] They found that the presence of 0.2 g of CO_2 (200 °C for 15 min) promoted the hydrolysis of xylan into low molecular weight sugars. In water-only reaction, the final xylose monomers yield from xylan was less than 5%, whereas, in CO_2-catalyzed reaction, a substantial xylose yield improvement was obtained. Additionally, they also reported that high-pressure CO_2–H_2O did not allow to produce degradation products in amounts as high as in case of HCl-catalyzed hydrolysis.

It is worth mentioning that different results were reported by McWilliams and van Walsum.[60] These authors compared the hydrolysis of aspen wood (instead of pure xylan) in high-pressure CO_2–H_2O to liquid hot water treatment over a temperature range of 180–220 °C. They found that the addition of CO_2 to water, and consequent formation of carbonic acid, neither improve the final xylose yields nor enhance production of furans.[60] Contrary to authors' statement, these results are not surprising, because aspen wood comprises of a high content of acetylated hemicelluloses. The liquid hot water promotes the cleavage of these acetylated hemicellulose bonds into high concentration of endogenous acid (*i.e.* acetic acid), which is strongly enough to promote catalytic effect without the need of carbonic acid formation. Therefore, this data could mean that the catalytic effect of high-pressure CO_2–H_2O is highly reliant on biomass composition and its recalcitrance. In addition to other factors, such as origin, age, climacteric conditions, harvesting method, *etc.*, physical properties of biomass are crucial factors in determination of the pre-treatment efficiency. Different biomass species, when subject to hydrolysis under similar reaction conditions, could give different sugar yields. For instance, hardwoods are generally characterized by higher density than that of softwoods, and this may hamper CO_2-catalyzed hydrolysis. However, some biomasses (*e.g.* switchgrass) are more accessible to hydrolysis and, independently of reaction

Table 5.2 Overview of high-pressure CO_2–H_2O hydrolysis of hemicelluloses derived from various lignocellulosic biomasses.[a]

Feedstock	Reaction conditions T (°C)	p_{CO_2} (bar)	t (min)	Reactor design	LSR (w/w)	Particle size (mm)	Total xylan sugars yield (g/100 g feedstock)	Xylooligomers yield (g/100 feedstock)	Ref.
Wheat straw	180	50	12	Batch	10	<1.5	88.6[c]	79.6[c]	54
Wheat straw	210	50	—[b]	Batch	10	<1.5	86.7[c]	70.6[c]	45
Wheat straw	215	30	—[b]	Batch	10	<1.5	76.7[c]	61.7[c]	46
Eucalyptus	160	50	80	Batch	10	<1	9.8	8.4	102
Corn stover	160	400	150	Batch	200	0.075–0.106	12.5[d]	—[e]	61
Corn cob	170	300	90	Batch	200	0.075–0.106	19.3[d]	—[e]	61
Switchgrass	170	450	60	Batch	200	0.075–0.106	13.2[d]	—[e]	61
Sugarcane bagasse	160	50	80	Batch	6.67	<1	15.8	12.6	64
Switchgrass	210/160[f]	200	1/60[f]	Batch	40[g]	0.038–1	5	8	103
Mixed hardwood	210/160[f]	200	16/60[f]	Batch	40[g]	0.038–1	2	10	103
Pure xylan	200	0.3[h]	15	Batch	100	—	15[d,i]	—[e]	59

[a]LSR: liquid-to-solid ratio; Total xylan sugar yield: sum of xylose monomers and xylooligomers.
[b]Reactor was cooled down when the desired temperature was attained.
[c]Yield express in g/100 of initial xylan content in feedstock.
[d]Only xylose yield (g/100 g feedstock) given.
[e]No data.
[f]Two-stage processs.
[g]Moisture content expressed in wt%.
[h]CO_2 added to the process (g).
[i]Estimated value based on a figure presented in the literature.

conditions, allow achieving higher sugar yields. King *et al.* studied high-pressure CO_2–H_2O hydrolysis of two different samples of biomass (switch-grass and corn stover).[61] With these different biomasses different total monomeric sugars were obtained. For instance, at 170 °C, 450 bar of CO_2 pressure for 60 min of reaction time the highest xylose yield from switch-grass was 13.2 wt%, whereas in the case of corn stover, the best xylose yield was 12.5 wt% but obtained under more severe reaction conditions (160 °C, 400 bar of CO_2 pressure for 150 min).[61] Due to differences in the structural and chemical composition of raw materials, the reaction conditions, mainly residence times, required to achieve the maximum monosaccharides yield also differ. This shows that biomass properties have a significant influence on the selection of experimental conditions and, consequently, on the effectiveness of pre-treatment. Additionally, van Walsum *et al.* reported that the formation of carbonic acid in reactional medium produced a hydrolysate with similar final pH in comparison to those obtained in liquid hot water experiments.[52] These surprising results can be explained by the low solubility of CO_2 at studied temperatures. Several literature studies reported the solubility of high-pressure CO_2 in water over a wide-range of temperatures and pressures.[40–42] It was reported that for temperatures up to 100 °C, the solubility of CO_2 in water decreases with an increase of temperature and increases with an increase of CO_2 pressure.[62] The relation between the pH value of the reactional medium and CO_2 pressure over a limited range of reaction temperatures has been calculated and presented by Chuang and Johanssen as depicted in Figure 5.4.[62]

Thus, based on the above consideration, high-pressure CO_2–H_2O should be performed at as low temperature as possible and high initial CO_2

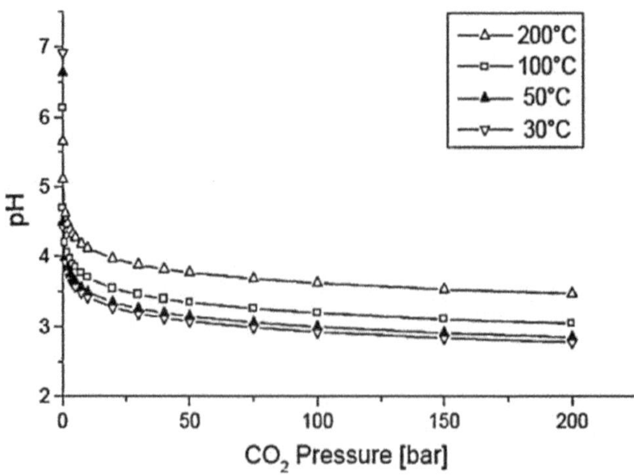

Figure 5.4 Influence of temperature and pressure of CO_2 on the pH value of the medium.
Reprinted from ref. 62. Copyright (2009) with permission from ISASF, International Society for Advancement of Supercritical Fluids.

pressures to promote high solubility of CO_2 in water and, consequently, more pronounced catalytic effect on polymers hydrolysis.

Moreover, the above-referred results can also be elucidated by the fact that pH was measured after reactions were finished. During the depressurization step a great part of CO_2 is removed; however, even after the CO_2 removal, the reactional mixture remains under non-equilibrium conditions because of low diffusivity of CO_2 in water. To achieve equilibrium, *i.e.* the situation when the concentration of CO_2 in the hydrolysate is equal to partial pressure in the atmosphere, an extended time and agitation are required.[45]

Besides aforementioned authors, Gurgel *et al.* reported the extraction of hemicellulose from sugarcane bagasse in high-pressure CO_2–H_2O.[63] The highest xylose concentration of 8.49 g L^{-1} was obtained at 115 °C and initial CO_2 pressure of 68 bar and within 60 min of reaction time. According to the authors, the obtained results indicate the potential of high-pressure CO_2–H_2O treatment in the reduction of requirements for further hydrolysis of C_5-oligosaccharides prior to fermentation processes.[63] Relvas *et al.* proposed kinetic models for the hydrolysis of hemicelluloses present in wheat straw under high-pressure CO_2–H_2O system at various initial CO_2 pressures (0 (liquid hot water), 20, 35 and 50 bar) at 180 °C and under isothermal conditions.[53] They found that the presence of CO_2 in reactional medium speeds up the initial reaction rate constant of hemicellulose hydrolysis into sugar oligomers, being the fastest step of hydrolytic process. The initial kinetic rate constant with water-only process is 40% lower than that observed for the lowest pressure (*e.g.* 20 bar of initial CO_2 pressure). Morais *et al.* studied the influence of carbonic acid formation on hydrolysis of hemicellulose wheat straw into xylose in both oligomer and monomer form over a range of reaction conditions (130, 215 and 225 °C and addition of 54 bar of initial CO_2 pressure).[46] They concluded that the addition of CO_2 adds value to the process by *in situ* formation of acid catalyst (*i.e.* carbonic acid) leading to higher production of hemicellulosic sugars in comparison to those obtain in CO_2-free processes under lower reaction temperature of 10 °C. High-pressure CO_2–H_2O outperformed autohydrolysis reaction leading to total sugar yield as high as 84% *vs.* 67.4% obtained in autohydrolysis processes.[46] Magalhães da Silva *et al.* demonstrated that the CO_2-catalyzed hydrolysis of wheat straw at relatively high temperatures (210 °C) and initial CO_2 pressure of 60 bar led to a total xylooligomers and xylose yield as high as 87% (g/100 g of initial xylan content).[45] Bogel-Łukasik and co-workers also studied the effect of high-pressure CO_2–H_2O over a wide range of initial CO_2 pressure (0, 20, 35 and 50 bar) and holding times (0 to 45 min) at reaction temperatures of 180 °C.[54] Similar to above-referred results, the presence of CO_2 enhanced the xylose oligomer yield leading to a maximum of 79.6 g per 100 g of initial xylan content (at 50 bar of initial CO_2 pressure and 12 min of residence time), whereas autohydrolysis process allowed to achieve 70.8 g per 100 g of initial xylan content. Additionally, higher initial CO_2 pressures decreases the oligosaccharides content being counterbalanced by the production of xylose monomers.[54] Zhang and Wu investigated the effect of supercritical CO_2 hydrolysis of sugarcane bagasse on the C_5-sugars yield.[64] The highest C_5-sugars yield of 15.8 g/100 g of feedstock, in which 80% corresponding to xylooligomers was obtained at 160 °C, 50 bar of CO_2 pressure for 80 min. Pang *et al.* compared various processes, such as H_2O_2, ammonia and H_2O_2, ammonia, NaOH, butanediol, 50% ethanol, liquid hot water, hot limewater and supercritical CO_2, in the pre-treatment of raw corn stalk.[65] Within the studied processes, supercritical CO_2 (150 °C, 80 bar of CO_2 pressure for 4 h) and butanediol (1,4-butanediol, 200 °C for 4 h) were the most effective in hemicellulose extraction (77%) from corn stalk.

5.3.3 Starch

Various biomasses contain starch as one of the main carbohydrates, and the hydrolytic processes require optimization aiming to obtain high yield of glucose, maltose and fructose in a favourable economically way. Unlike cellulose polymer, starch is easily hydrolyzed in hydrothermal processes without the addition of catalysts. Mineral-acid catalyzed hydrolysis of starch into its corresponding sugars is an old industrial process. Nowadays, there is an increasing need to replace these types of technologies for more ecological ones such as CO_2-catalyzed processes (Table 5.3). Miyazawa and Funazukuri reported that the addition of 0.3 g of CO_2 (at 200 °C for 15 min) to hydrolysis of starch derived from sweet potato led to glucose yield 14-fold higher than those obtained in water alone.[59] Brunner found that high-pressure CO_2–H_2O hydrolysis (230 °C, 240 bar of CO_2 pressure in a tubular reactor with reaction times of about 180 s) of corn starch improved the glucose yield from 5% (CO_2-free process) up to 60%.[66] Figure 5.5 shows the influence of CO_2 addition and respective concentration in glucose production from corn starch.

Moreschi *et al.* studied the hydrolysis of ginger bagasse using subcritical water and CO_2 to produce reducing sugars and other low molecular-weight products, over a wide range of temperatures (176, 188 and 200 °C) within 150 bar of CO_2 pressure.[67] The highest hydrolysis degree of 97.1% (after 15 min) and the maximum reducing sugars yield of 18.1% (after 11 min) were obtained for the highest reaction temperature (200 °C). Additionally, they characterized the reaction products obtained from high-pressure CO_2–H_2O hydrolysis of ginger bagasse. Products with molecular weight between 10 000–100 000 Da constituting 30–40% and 60–75% of the total products found at 176 and 188 °C, respectively, whereas at 200 °C they contributed to only 2.5–10% of all products. The increase of temperature to 200 °C led to formation of smaller oligosaccharides, which reached 100% of total product within 15 min of reaction time. Raffinose (trisaccharide) was identified prior to 5 min of reaction time and after that appeared to be converted into sucrose and glucose, whereas the absence of glucose confirms its conversion into fructose.[67] The same authors also stated that temperature favours the occurrence of undesired reactions harming the high yield of reducing sugars. Thus, based on the aforementioned data, the hydrolysis of starchy

Table 5.3 Overview of high-pressure CO_2–H_2O hydrolysis of starch-based biomasses.[a]

Feedstock	Reaction conditions T (°C)	p_{CO_2} (bar)	t (min)	Reactor design	LSR (w/w)	Particle size (mm)	Glucose/reducing sugar yield (g/100 g starch)	Other results	Ref.
Pure starch from sweet potato	200	0.3[b]	15	Batch	100	—[c]	53[d]	Yield of starch without CO_2 addition: <5%; Fructose and levoglucan yields: 4.7 and 2.7%, respectively; Oligosaccharides yield: 34%; Total sugar yield: 88%[e]	59
Pure starch from sweet potato	200	30	15	Batch	25	—[c]	54.8	Production of 5-HMF	69
Corn starch	230	—[f]	3.5	Continuous	1[g]	—[c]	>60[d]	—	55
Ginger (*Zingiber officinale Roscoe*) bagasse	200	150	11	Batch	2.3	—[c]	18.1	Hydrolysis degree: 97.1% (after 15 min of reaction)	67
Ginger (*Zingiber officinale Roscoe*) bagasse	188	150	1	Batch	2.3	—[c]	—	Raffinose and fructose in liquors: 3.0 wt/v % and 0.7 wt/v %	67
Dried ginger (*Zingiber officinale Roscoe*)	200	150	11	Batch	2.3	3.0	7.1	Total sugar yield: 43%	68
Dried turmeric (*Curcuma longa* L.)	200	150	11	Batch	2.3	3.0	10.6	Total sugar yield: 49%	68

[a]LSR: liquid-to-solid ratio.
[b]CO_2 added to the process (g).
[c]No data.
[d]Glucose yield.
[e]Sum of monomers and oligosaccharides of glucose and fructose.
[f]CO_2-saturated water corresponding to 82.1 mg of CO_2/g H_2O.
[g]Biomass loading expressed in wt%.

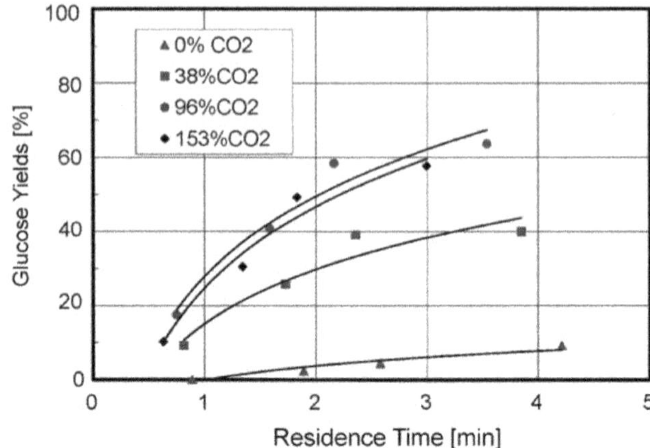

Figure 5.5 Effect of CO_2 addition on glucose yield from corn starch. Concentration of CO_2 is expressed in percentage of saturation.
Adapted from G. Brunner, Near critical and supercritical water. Part I. Hydrolytic and hydrothermal processes, *J. Supercrit. Fluids*, **47**, 373–381,[66] Copyright (2009) with permission from Elsevier.

biomass should be carried out at as low temperature as possible and simultaneously under high pressures of CO_2 to enhance the solubility of CO_2 in water. The same research group reported that the application of high-pressure CO_2 to subcritical water improved the final degree of hydrolysis of fresh and dried turmeric (*Curcuma longa L.*) and ginger (*Zingiber officinale R.*).[68] The highest hydrolysis of 97%—98% were found for both ginger and fresh turmeric. The highest total sugar yield (74%) was obtained for ginger bagasse at 200 °C, 150 bar of CO_2 pressure for 11 min of reaction time.[68] Orozco *et al.* aimed to investigate the effectiveness and viability of fermentable hydrolysate production from starch using the high-pressure CO_2–H_2O process.[69] Starch from potato powder was subject to hydrolysis under various initial concentrations (40, 120 and 200 $g L^{-1}$) in presence of CO_2 (30 bar) and water at 180–235 °C for 15 min. As expected, the incorporation of CO_2 in the water experiments improved the glucose yield reaching a maximum of 548 $g kg^{-1}$ of starch (40 $g L^{-1}$ initial starch concentration), accompanied by the formation of undesired products, *i.e.* 5-HMF.[69]

Combined technologies (*e.g.* microwave) exhibit a synergetic effect on starch hydrolysis. This could be considered an option in the field of starch conversion. Research about the effect of high-pressure CO_2, coupled with aforementioned microwave technologies with the objective to increase the starch sugars yield without formation of degradation products, was carried out by Thangavelu *et al.*[20] They found that microwave hydrothermal hydrolysis of sago pith waste was enhanced when CO_2 (dry ice) was added to the process. The highest glucose yield of 43.8% was obtained when CO_2 was added in microwaves assisted process performed at 900 W of 2 min of irradiation.[20]

5.3.4 Proteins

The depolymerization of proteins has attracted importance due to environmental and economic reasons. The conversion of proteins towards amino acids has been extensively investigated using hydrothermal technologies.[34,70] Unexpectedly, a very limited number of protein hydrolysis experiments in presence of CO_2 have been reported. The addition of hydronium ion originated from carbonic acid to nitrogen of the peptide bond is the determining step and because of this the formation of carbonic acid is a factor that determines the hydrolysis reaction. Scheme 5.4 depicts the mechanism of acid-catalyzed hydrolysis of peptides.

Rogalinski *et al.* studied the effect of CO_2 addition to subcritical water on hydrolysis of protein (BSA, an elliptic protein molecule containing of a chain of 607 amino acids) for different levels of CO_2 saturation (temperature of 250 °C was kept constant) and residence times as shown in Figure 5.6.[34] The obtained results allow to conclude that in CO_2-free subcritical water process, the concentration of amino acids was 36.6 ($mg_{amino\ acids}\ (g_{BSA})^{-1}$) for residence time of 300 s, whereas for CO_2 present at the highest level of

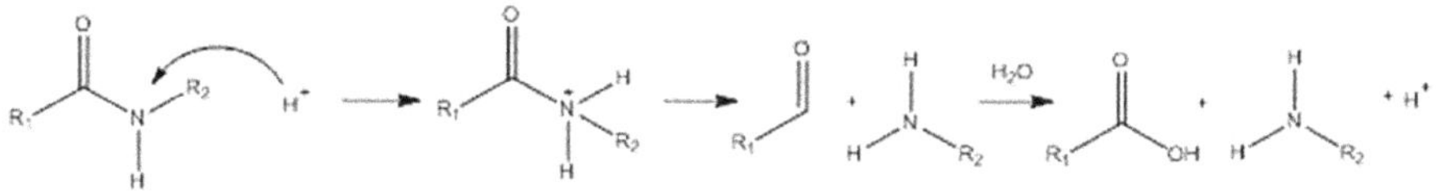

Scheme 5.4 Reaction scheme of acid-catalyzed hydrolysis of peptide.

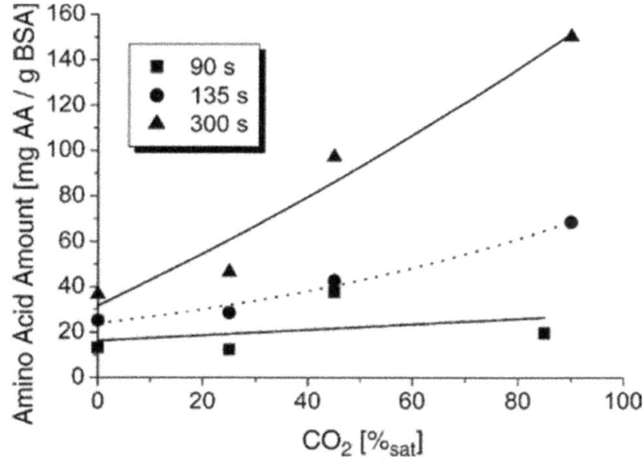

Figure 5.6 Effect of CO_2 saturation on production of amino acids from BSA registered for various reaction times. Temperature of 250 °C was kept constant in all experiments.
Reprinted from T. Rogalinski, S. Herrmann and G. Brunner, Production of amino acids from bovine serum albumin by continuous sub-critical water hydrolysis, *J. Supercrit. Fluids*, **36**, 49–58,[34] Copyright (2005) with permission from Elsevier.

CO_2 saturation the concentration of amino acids increased four times to 150.3 $mg_{amino\ acids}\ (g_{BSA})^{-1}$. These data demonstrated that presence of CO_2 favours the acid hydrolysis of proteins, without the need of supplementary acids.

5.3.5 Lignin

5.3.5.1 *Delignification*

Lignin is one of the most promising fractions to produce valuable chemicals, such as guaiacol, phenol, cresol, vanillin, tricin, *etc.*, as well as high-value renewable polymers manufactured by polycondensation,[71] ring opening polymerization[72] or electrophilic aromatic substitution.[73] Thus, it is utmost importance to find out and to develop more sustainable and cost-effective technologies able improve the biorefinery economics by the production of high-value biomaterials and chemicals from lignin. In delignification processes, an organic solvent is added, usually together with water and frequently with an acid catalyst. Water acts as nucleophile and interacts with centres previously activated by the acid catalyst in the proto-lignins (*i.e.* "*in situ* lignin"). Organic solvent, in turn, dissolves lignin and contributes to the impregnation of lignocellulosic biomass.[74,75] An effective alteration in the organosolv process, which could improve the delignification yield, regardless of type of lignocellulosic biomass, is highly dependent on addition of high-pressure fluid to the process and on process conditions.[74,75] Therefore, this section is focused on the use of high-pressure CO_2 as co-solvent, to separate lignin from the biomass matrix without degradation of other biomass key components.

The addition of CO_2 as solvent for biomass delignification shows several advantages, *e.g.* formation of anomalous porosity and lamellar structure of lignocellulosic biomass, combined with an increased diffusivity of solvents.[75] Thus, biomass is more susceptible to organic solvents (*e.g.* dioxane, acetic acid, ethanol, methanol, *etc.*) leading to more efficient delignification. An overview of results reported in literature related to the use of high-pressure CO_2, as solvent in biomass delignification is summarized in Table 5.4.

Machado *et al.* studied the removal of lignin from *Eucalyptus globulus* in a presence of pure 1,4-dioxane and high-pressure CO_2–1,4-dioxane binary systems at 170 bar of pressure for 300 min over the temperature range of 160–180 °C.[76] At 170 °C and in pure 1,4-dioxane with flow rate of 2.1 $g\,min^{-1}$, the maximum removal of lignin obtained was 75%. However, under these conditions, a great removal of hemicelluloses (60%) was observed, whereas the most pronounced mass loss related to cellulose (close to 70%) was found at 180 °C. The CO_2–1,4-dioxane binary system was investigated at 170 °C, 170 bar within 300 min of residence time. The dioxane flow rate was 1.38 $g\,min^{-1}$ and the CO_2 flow rate was adjusted from 0.58–5.23 $g\,min^{-1}$ to obtain different solvent system composition.

Table 5.4 High-pressure CO_2-mediated delignification of various lignocellulosic biomasses.

Feedstock	Reaction conditions T (°C)	p (bar)	t (min)	Solvents	Pulp yield (%)	Delignification yield (%)	Other results/comments	Ref.
Sugarcane bagasse	190	160	60	CO_2 and C_2H_6O–H_2O, 3 : 1 v/v	32.7	88.4	Non-selective; high mass losses	75
Sugarcane bagasse[a]	180	250	120	CO_2 and C_2H_6O–IL[b]; 20 : 1 v/v	—	41.9	Highly selective for delignification	82
Sugarcane bagasse	190	70	105	CO_2 and $C_4H_{10}O$–H_2O; 6 : 4 v/v	8.7	94.5	Higher pressures > pulp yield; Higher temperature < pulp and polyoses yield	104
Pinus taeda wood chips	190	160	60	CO_2 and C_2H_6O–H_2O, 1 : 1 v/v	43.7	93.1	Non-selective; high mass losses	75
Corn stover	200	130	80	CO_2 and C_2H_6O–H_2O; 2 : 1 v/v	—	90.1	Production of phenolic compounds, such as vanillin, 4-hydroxy-3-methoxystyrene, and 2,3,6-trimethyl-4-methoxy-phenol were identified at 180 °C, 130 bar and 60 min	79
Aspen chips	140	138	240	CO_2–SO_2, 98 : 2, mol/mol	—	84	—	81
Red spruce	180	250	180	CO_2–CH_3COOH–H_2O, 11 : 65 : 24, mol/mol/mol	—	89	The highest delignification for pure CH_3COOH: 93%	77
Eucalyptus globulus wood	170	170	300	CO_2–$C_4H_8O_2$; ~30 : 70 w/w[c]	—	~40[c]	The highest delignification yield for pure $C_4H_8O_2$: 75%;	76
Miscanthus X giganteus	200	55	60	CO_2 and H_2O–C_2H_6O, 1 : 1, v/v	—	86	Dissolution: 53%	80

[a]Feedstock was previously subject to extraction with ethanol to remove most of ethanol-extractable components.
[b]1-Butyl-3-methylimidazolium acetate (95%).
[c]Data estimated from figure.

The highest delignification yield, in presence of CO_2–1,4-dioxane binary system (~30 : 70), was close to 40%, whereas, for hemicelluloses, the removal reached 100% with 54.5% of 1,4-dioxane content in the mixture. The mass loss of cellulose was kept below 11% for all studied solvent compositions. These low delignification yields can be attributed to absence of nucleophilic solvents in the reactional medium to promote the cleavage of ether bonds between lignin and carbohydrates and the capacity to dissolve lignin fragments.[76] Thus, it can be concluded that to improve the delignification yield, the addition of water to the reactional medium could be required. Kiran and Balkan investigated the delignification of red spruce wood performed in a presence of acetic acid–water, CO_2 and acetic acid, and CO_2 and acetic acid–water mixtures.[77] The effect of composition of acetic acid–water binary mixture was studied from 14 to 250 bar of pressure over a wide range of temperatures (120–180 °C) and reaction times (30–180 min). In the presence of CO_2, the highest delignification yield of 89% was obtained in CO_2–acetic acid–water mixture composition of 11 : 65 : 24 (mole fraction) under 180 °C and 180 min of residence time. The best delignification yield of 93% was obtained in CO_2-free process, *i.e.* in presence of acetic acid–water mixture composition of 73 : 27 (mole fraction) under above-referred reaction conditions.[77] Another work from the same group, demonstrated the effect of different binary (CO_2–ethanol, CO_2–water and ethanol–water) and ternary (CO_2–water–ethanol) systems on the delignification of red spruce wood.[78] The CO_2–ethanol (0.76 CO_2 mole fraction) mixture under high-pressure conditions (167 °C and 325 bar of pressure) was employed and 4.8% of dissolution was obtained after 120 min of reaction time. Two different compositions of CO_2–water mixture (0.39 and 0.88 CO_2 mole fraction) at 190–193 °C and 290 bar yielded for 120 min yielded 17.5% and 35.9% of higher and lower CO_2 content, respectively. Finally, two different ethanol/water compositions (0.91 and 0.52 water mole fraction) were studied at 190 °C and 290 bar of pressure and allowed to achieve 41.4% and 32.4% at higher and lower water content, respectively. The CO_2–water–ethanol (0.96 and 0.022 mole fraction of CO_2 and H_2O, respectively) system was employed under the same conditions as those aforementioned for ethanol–water system and 19.3% of dissolution yield of red spruce wood was obtained.[78] Pasquini *et al.* described the delignification of *Pinus taeda* wood chips and sugarcane bagasse with CO_2–ethanol–water at different temperatures (142–198 °C), pressures (145–230 bar) and reaction times (30–150 and 30–120 for *P. taeda* wood and sugarcane bagasse, respectively).[75] The delignification yields were in the range of 88–93%, whereas the pulping yields were 43.7% and 32.7% for *P. taeda* wood and sugarcane bagasse, respectively, at 190 °C, 160 bar and 60 min. Lv *et al.* also studied the process of delignification of corn stover in high-pressure CO_2 as co-solvent for ethanol–water mixture.[79] An L_9 (3^3) set of orthogonal experiments was used to investigate the effects of reaction temperature, pressure and residence time in rate of delignification. It was found that reactional medium consisting of CO_2–ethanol–water can effectively extract lignin from corn stover. Among tested variables, temperature was the most influencing factor followed by reaction time and pressure. The highest delignification rate of 90.1% was obtained at 200 °C, 130 bar within 80 min of residence time.[79] The morphological changes in corn stover after treatment with CO_2–ethanol–water were assessed by scanning electron microscopy (SEM). Figure 5.7 illustrated the structure of native corn stover and morphological changes of corn stover after process. Although the authors did not comment on this comparing SEM images of native corn stover to the insoluble fibres of corn stover after treatment it is notorious that fibres have been preserved as the majority of the lignin has been extracted. Thus, CO_2–ethanol–water, at above-referred reaction conditions, promotes the delignification of biomass without destroying the cellulose fibres. Also Roque *et al.* optimized the delignification of *Miscanthus X giganteus* using water–CO_2–ethanol mixture.[80] The maximum of 86% of delignification and 53% of dissolution were obtained in reaction mixture composed by water and ethanol in 1 : 1, v/v at 200 °C, 55 bar of initial CO_2 pressure for 60 min of residence time. Analogously to Lv's group, Roque *et al.* found that water–CO_2–ethanol mixture, under optimal reaction conditions, promotes delignification preserving the cellulose polymer.

Shah *et al.* investigated the effect of CO_2–SO_2 (98 : 2, mol%/mol%) on the removal of lignin from carbohydrates in hardwood.[81] The highest lignin removal yield of 84% was obtained for aspen chips at 140 °C and for 4 h. They also reported a high loss of carbohydrates for temperatures above 130 °C. Silveira *et al.*, for the first time, reported a new process of the

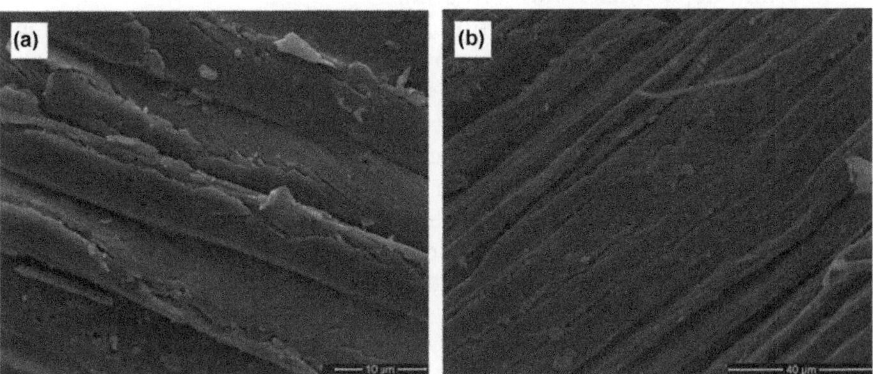

Figure 5.7 Scanning electron microscopy of native corn stover (a) and treated in presence of CO_2–ethanol–water at 180 °C, 130 bar for 60 min (b). Adapted from H. S. Lv, L. Yan, M. H. Zhang, Z. F. Geng, M. M. Ren and Y. P. Sun, Influence of Supercritical CO_2 Pretreatment of Corn Stover with Ethanol-Water as Co-Solvent on Lignin Degradation, *Chem. Eng. Technol.*, **36**, 1899–1906,[79] Copyright (2005) with permission from John Wiley & Sons.

sugarcane bagasse delignification based on the use of ionic liquid (IL), supercritical CO_2 and ethanol, called "IL-assisted supercritical organosolv pre-treatment".[82] The IL used was 1-butyl-3-methylimidazolium acetate, [bmim][OAc], because acetate-containing ILs are capable to induce cellulose amorphogenesis and provide its partial dissolution.[82,83] The treatment resulted in a 41.9% of delignification of sugarcane bagasse and total mass recovery of 87% (180 °C, 250 bar for 120 min), which indicates that this particular IL-assisted supercritical organosolv pre-treatment is quite selective for delignification.[82]

Concluding it can be stated that the composition of reactional mixtures, in particular the presence of CO_2, has a great influence on the efficiency of delignification. The efficiency of delignification is important when one of aims is to produce valuable chemicals from lignocellulosic biomass, since this would allow still using carbohydrate polymers present in the material. These carbohydrates can be further converted into their sugar constituents and these through fermentation produce biofuels, dicarboxylic acids, *etc.*, while lignin can be converted into added-value compounds, such as aromatic hydrocarbons for fuel and chemical applications.

5.3.5.2 *Depolymerization*

Current strategies to produced valuable compounds from lignin are typically based on alkaline[84,85] or acid hydrolysis,[86] hydrogenation[87] or flash pyrolysis depolymerization[88,89] into a mixture of aromatic compounds. These compounds can also be further converted by hydrolysis of methoxy groups. The main lignin reaction pattern is strongly dependent on the chemical structure of the lignin used.[86] Scheme 5.5 illustrates an example of hydrothermal depolymerization of lignin.[90]

One of the most common lignin depolymerization approach involves the use of sub/supercritical water, as reactional medium, because with the increase of reaction temperature the changes in the physicochemical properties of H_2O (dielectric constant and ion product) indicate that H_2O at high temperatures, along organic solvents, is more appropriate for depolymerization reaction.[91] However, the reaction conditions needed to achieve supercritical H_2O are energy demanding and the use of catalysts can be also an expensive option. Thus, the depolymerization of lignin carried out in presence of CO_2 and subcritical H_2O has been considered an interesting alternative because it might increase the yield and selectivity towards the formation of specific phenolic compounds, without the need to maintain supercritical conditions and catalyst addition. Recently, Numan-Al-Mobin investigated the effect of high-temperature water (200–500 °C) with CO_2 or N_2 (used as standard system with pressure equal to this of CO_2) at 220.63 bar and fixed reaction time of 10 min on the production valuable commodities from alkali lignin.[92] Depending on the reaction conditions, different phenolic compounds, such as guaiacol and its homologous were produced, as depicted in Figure 5.8.

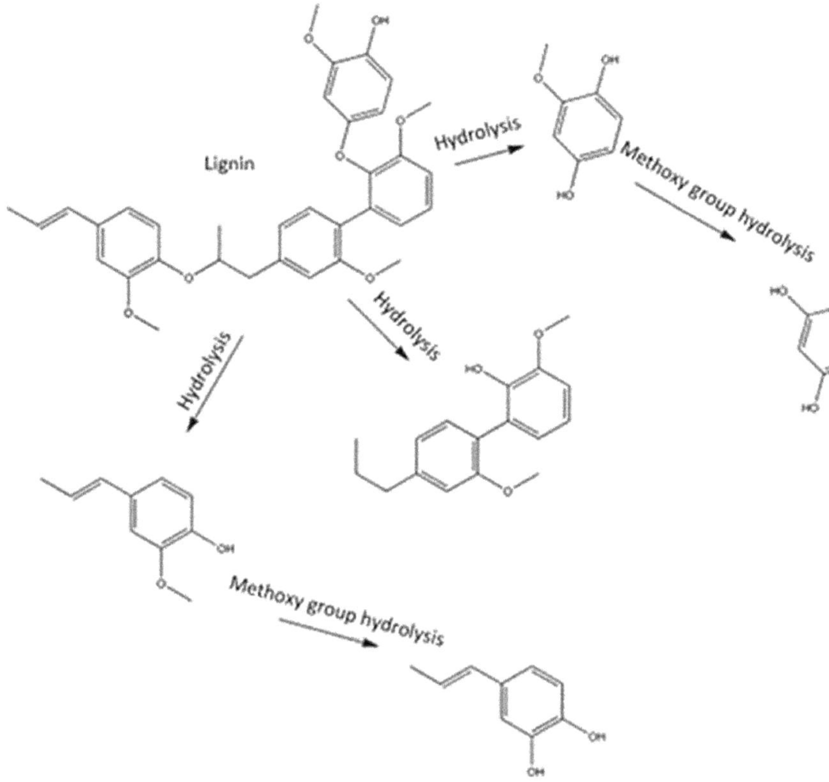

Scheme 5.5 Simplified scheme of lignin depolymerization.
Adapted with permission from A. Liu, Y. Park, Z. L. Huang, B. W. Wang, R. O. Ankumah and P. K. Biswas, *Energy Fuels*, 2006, **20**, 446–454.[90] Copyright (2006) American Chemical Society.

More pronounced differences between CO_2-and N_2-assisted hydrothermal process were obtained at 300 °C. High relative yields (60%) of guaiacols were found in presence of CO_2, whereas in presence of N_2 the yields were only 37%. It was stated that as long as CO_2 acts as acidic homogeneous catalysts it can influence the pattern of lignin depolymerization. Guaiacyl acids were the second most abundant product in CO_2-assisted process, however they were found in higher yields in presence of N_2. This study also illustrates the influence of temperature regarding the efficiency of catalyst. In a presence of CO_2, an increase of temperatures (from 300 to 400 °C) decreases the CO_2 solubility and, consequently, the catalyst-enabled selectivity. However, high temperatures promote the formation of other phenolic derivatives (*e.g.* phenols), suggesting lower reaction selectivity.[92] Regarding the resulting materials obtained with CO_2-assisted hydrothermal, at 250 and 350 °C, they were analyzed by pyrolysis – GC – MS. The obtained data demonstrated a great content of phenolic compounds, which were not detected either in untreated lignin or in solid materials obtained at lower temperatures (Figure 5.9),

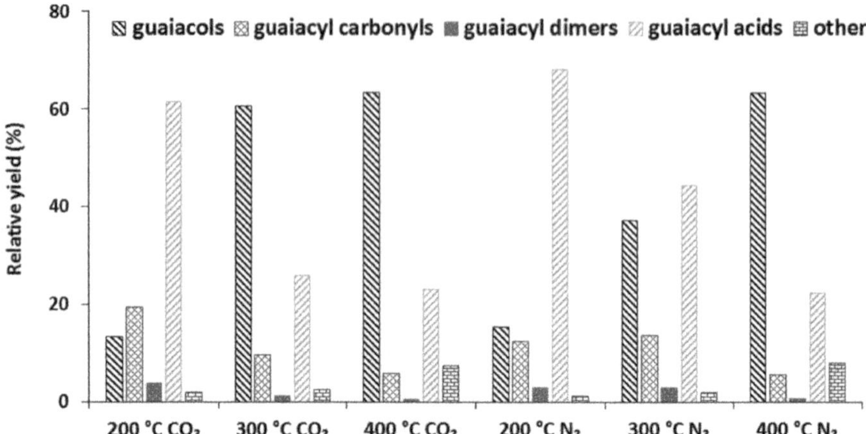

Figure 5.8 Phenolic compounds produced after hydrothermal processing, in presence of CO_2 or N_2, at different temperatures (200, 300 and 400 °C) and fixed pressure of 220.63 bar and for 11 min.
Reprinted with permission from A. M. Numan-Al-Mobin, K. Voeller, H. Bilek, E. Kozliak, A. Kubatova, D. Raynie, D. Dixon and A. Smirnova, *Energy Fuels*, 2016, **30**, 2137–2143.[92] Copyright (2016) American Chemical Society.

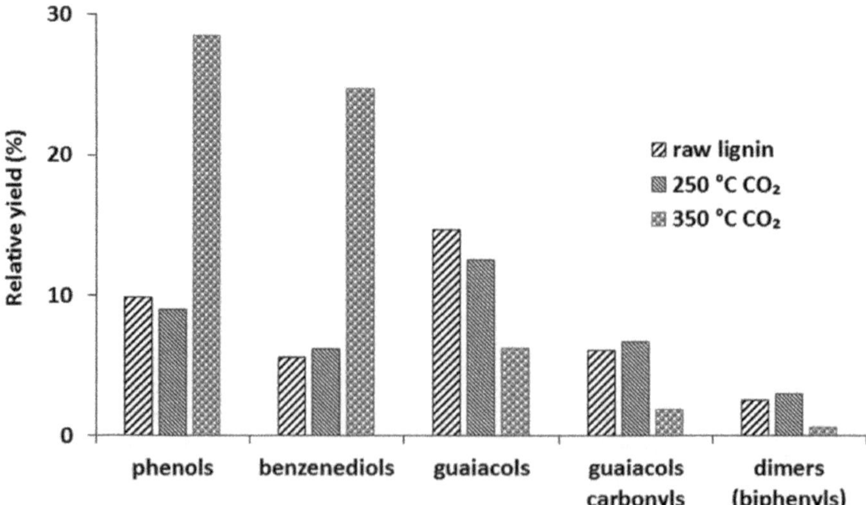

Figure 5.9 Pyrolysis – GC – MS results for untreated lignin (raw lignin) and for resulting materials obtained in CO_2-assisted hydrothermal process at 250 and 350 °C and fixed pressure of 220.63 bar for 11 min.
Reprinted with permission from A. M. Numan-Al-Mobin, K. Voeller, H. Bilek, E. Kozliak, A. Kubatova, D. Raynie, D. Dixon and A. Smirnova, *Energy Fuels*, 2016, **30**, 2137–2143.[92] Copyright (2016) American Chemical Society.

indicating that these materials can be used as a source of valuable phenolics. The authors concluded that the addition of CO_2 to hydrothermal processes has a great importance since it could increase the selective production of phenolic compounds and later they can be use in the production of polymers with desirable properties.[92]

The deconstruction of lignin may also be carried out under milder conditions in presence of CO_2, water and ethanol as solvents. Schrems *et al.* performed several studies on the combined supercritical CO_2 with water–ethanol mixture in the pulping process using lignin model compounds.[93] The proposed reactions were conducted at reasonable, for lignin processing, temperatures of 140, 160 and 180 °C and 60 bar of initial pressure. The presence of high-pressure CO_2 increased the effectiveness of delignification process by reduction of activation energy and by accelerating the lignin depolymerization. For instance, the lignin decomposition rate at 140 °C with high-pressure CO_2 was faster than the reaction rate at 180 °C in the absence of CO_2. This trend might be explained by decrease of the solvent polarity that promotes the neutral *para*-quinone methide intermediate and by destabilization of the ionic intermediates and polar transition states. Although the presence of CO_2 promotes the decomposition of model lignin compounds, surprisingly it also slows down the formation of stable end products even at low temperature. Additionally, Schrems *et al.* stated that the presence of CO_2 not only changed the polarity, diffusivity and accessibility in the pulping system but also modified the chemical pathways and energy activation parameters and, consequently the kinetics of delignification process.[93] Recent findings in deconstruction of lignin present in corn stover using CO_2 and water–ethanol (2 : 1 v/v) at 180 °C, 130 bar for 60 min have been reported.[79] The proposed aqueous phase methodology generated mainly esters, aldehydes, ketones and acids, and phenolic products, such as 4-hydroxy-3-methoxystyrene and vanillin. This work also highlights the influence of CO_2 and water–ethanol system and reaction conditions with respect to degradation of hemicellulose polymer. Lv *et al.* found high quantities of furfural and acetic acid in the aqueous phase indicating the extensive hemicellulose degradation at relatively mild process conditions however, low quantities of aromatic and significant degradation of hemicellulose are the main limitations of the proposed technology. Gosseling *et al.* also performed several studies on the combination of supercritical CO_2 with acetone/water mixture on depolymerization of organosolv hardwood and wheat straw lignins at high temperatures (300–370 °C), 100 bar of pressure for 3.5 h.[94] Under the reported conditions, lignins were converted into monomeric aromatic compounds (10–12%) in presence of formic acid (14 wt% based on lignin) as hydrogen donor aiming the stabilization of aromatic radicals. Hardwood and wheat straw lignins yielded different mixture of aromatic compounds, with the highest individual yield of 3.6% for syringol and 2.0% for syringic acid based on lignin, respectively. However, high formation of char (45–47%) was found for both lignins.

5.4 Conclusions

High-pressure CO_2 and CO_2–H_2O mixtures have gained a growing interest as solvents as suitable technologies for a direct hydrolysis of biomass polymers. These technologies are considered as eco-friendly processes, since they do not use hazardous chemicals and do not require waste treatments. Furthermore, these technologies integrate the strengths of various biomass technologies, such as mechanical pre-treatments (*e.g.* CO_2 explosion), chemical pre-treatment (liquid hot water or dilute-acid hydrolysis) in a single technology. High-pressure CO_2 is able to penetrate deeper into small pores of lignocellulosic biomass structure than other solvents promoting its disruption; when dissolved in water (which equally can be the biomass moisture) CO_2 makes the reactional medium more acidity turning the biomass processing into a process similar to dilute-acid hydrolysis. High-pressure CO_2 and CO_2–H_2O showed to be very effective for various types of lignocellulosic biomasses, including hardwoods and softwoods. However, highly acetylated biomasses seem to impair the performance of high-pressure CO_2 as catalyst.

A lot of research on hemicellulose polymers conversion has been done. Among them are works about a selective hydrolysis of hemicelluloses into its components leaving the conversion of other polymers and the influence of biomass endogenous acids production (*i.e.* acetic acid) almost unexplored.

Acknowledgements

This work was supported by the Fundação para a Ciência e Tecnologia (FCT, Portugal) through strategic project UId/QUI/5006/2013, and grants SFRH/BD/94297/2013 (ARCM) and IF/00424/2013 (RML).

References

1. P. N. R. Vennestrom, C. M. Osmundsen, C. H. Christensen and E. Taarning, *Angew. Chem., Int. Ed.*, 2011, **50**, 10502–10509.
2. J. H. Clark, F. E. I. Deswarte and T. J. Farmer, *Biofuels, Bioprod. Biorefin.*, 2009, **3**, 72–90.
3. IPCC, *IPCC Special Report on Renewable Energy Sources and Climate Change Mitigation*, IPCC, Cambridge, UK, 2011.
4. A. Alaswad, M. Dassisti, T. Prescott and A. G. Olabi, *Renewable Sustainable Energy Rev.*, 2015, **51**, 1446–1460.
5. A. Corma, S. Iborra and A. Velty, *Chem. Rev.*, 2007, **107**, 2411–2502.
6. P. Gallezot, *Chem. Soc. Rev.*, 2012, **41**, 1538–1558.
7. R. A. Sheldon, *Green Chem.*, 2014, **16**, 950–963.
8. R. A. Sheldon, *J. Mol. Catal. A: Chem*, 2016, **422**, 3–12.
9. C.-H. Zhou, X. Xia, C.-X. Lin, D.-S. Tong and J. Beltramini, *Chem. Soc. Rev.*, 2011, **40**, 5588–5617.

10. M. H. L. Silveira, A. R. C. Morais, A. M. da Costa Lopes, D. N. Olekszyszen, R. Bogel-Lukasik, J. Andreaus and L. P. Ramos, *ChemSusChem*, 2015, **8**, 3366–3390.
11. Y. Sun and J. Y. Cheng, *Bioresour. Technol.*, 2002, **83**, 1–11.
12. M. J. Taherzadeh and K. Karimi, *Int. J. Mol. Sci.*, 2008, **9**, 1621–1651.
13. F. H. Isikgor and C. R. Becer, *Polym. Chem.*, 2015, **6**, 4497–4559.
14. V. B. Agbor, N. Cicek, R. Sparling, A. Berlin and D. B. Levin, *Biotechnol. Adv.*, 2011, **29**, 675–685.
15. M. E. Himmel, S. Y. Ding, D. K. Johnson, W. S. Adney, M. R. Nimlos, J. W. Brady and T. D. Foust, *Science*, 2007, **315**, 804–807.
16. N. Mosier, C. Wyman, B. Dale, R. Elander, Y. Y. Lee, M. Holtzapple and M. Ladisch, *Bioresour. Technol.*, 2005, **96**, 673–686.
17. H. Roper, *Starch-Starke*, 2002, **54**, 89–99.
18. H. Jorgensen, J. B. Kristensen and C. Felby, *Biofuels Bioprod. Biorefin.*, 2007, **1**, 119–134.
19. P. Kumar, D. M. Barrett, M. J. Delwiche and P. Stroeve, *Ind. Eng. Chem. Res.*, 2009, **48**, 3713–3729.
20. S. K. Thangavelu, A. S. Ahmed and F. N. Ani, *Appl. Energy*, 2014, **128**, 277–283.
21. R. Rinaldi and F. Schuth, *ChemSusChem*, 2009, **2**, 1096–1107.
22. L. Laureano-Perez, F. Teymouri, H. Alizadeh and B. E. Dale, *Appl. Biochem. Biotechnol.*, 2005, **121**, 1081–1099.
23. D. Fengel and G. Wegener, *Wood: Chemistry, Ultrastructure, Reactions*, de Gruyter, New York, NY, USA, 1983.
24. B. C. Saha, *J. Ind. Microbiol. Biotechnol.*, 2003, **30**, 279–291.
25. H. V. Scheller and P. Ulvskov, *Plant Biol.*, 2010, **61**, 263.
26. O. Bobleter, *Prog. Polym. Sci.*, 1994, **19**, 797–841.
27. O. Faix, *Holzforschung*, 1991, **45**, 21–27.
28. M. Kleinert and T. Barth, *Chem. Eng. Technol.*, 2008, **31**, 736–745.
29. E. M. Rubin, *Nature*, 2008, **454**, 841–845.
30. C. P. Xu, R. A. D. Arancon, J. Labidi and R. Luque, *Chem. Soc. Rev.*, 2014, **43**, 7485–7500.
31. C. Chatterjee, F. Pong and A. Sen, *Green Chem.*, 2015, **17**, 40–71.
32. W. L. Bao, D. M. Omalley and R. R. Sederoff, *Proc. Natl. Acad. Sci. U. S. A.*, 1992, **89**, 6604–6608.
33. I. Chibata and K. Kawashima, in *Nutrition: Proteins and Amino Acids*, ed. T. Suzuki, A. Yoshida and H. Naito, Springer Verlag, 1990, pp. 273–284.
34. T. Rogalinski, S. Herrmann and G. Brunner, *J. Supercrit. Fluids*, 2005, **36**, 49–58.
35. E. Scott, F. Peter and J. Sanders, *Appl. Microbiol. Biotechnol.*, 2007, **75**, 751–762.
36. A. A. Peterson, F. Vogel, R. P. Lachance, M. Froling, M. J. Antal and J. W. Tester, *Energy Environ. Sci.*, 2008, **1**, 32–65.
37. J. S. Luterbacher, J. W. Tester and L. P. Walker, *Biotechnol. Bioeng.*, 2010, **107**, 451–460.

38. M. Stamenic, I. Zizovic, R. Eggers, P. Jaeger, H. Heinrich, E. Roj, J. Ivanovic and D. Skala, *J. Supercrit. Fluids*, 2010, **52**, 125–133.
39. K. L. Toews, R. M. Shroll, C. M. Wai and N. G. Smart, *Anal. Chem.*, 1995, **67**, 4040–4043.
40. H. Teng and A. Yamasaki, *J. Chem. Eng. Data*, 1998, **43**, 2–5.
41. R. Wiebe and V. L. Gaddy, *J. Am. Chem. Soc.*, 1940, **62**, 815–817.
42. A. N. Sabirzyanov, A. P. Il'in, A. R. Akhunov and F. M. Gumerov, *High Temp.*, 2002, **40**, 203–206.
43. P. B. Stewart and P. Munjal, *J. Chem. Eng. Data*, 1970, **15**, 67–71.
44. J. W. King and K. Srinivas, *J. Supercrit. Fluids*, 2009, **47**, 598–610.
45. S. P. Magalhães da Silva, A. R. C. Morais and R. Bogel-Lukasik, *Green Chem.*, 2014, **16**, 238–246.
46. A. R. C. Morais, A. C. Mata and R. Bogel-Lukasik, *Green Chem.*, 2014, **16**, 4312–4322.
47. G. P. van Walsum, *Appl. Biochem. Biotechnol.*, 2001, **91–93**, 317–329.
48. R. P. Overend, E. Chornet and J. A. Gascoigne, *Philos. Trans. R. Soc., A*, 1987, **321**, 523–536.
49. S. Miller and R. Hester, *Chem. Eng. Commun.*, 2007, **194**, 85–102.
50. S. I. Mussatto, G. Dragone, M. Fernandes, A. M. F. Milagres and I. C. Roberto, *Cellulose*, 2008, **15**, 711–721.
51. F. M. Girio, C. Fonseca, F. Carvalheiro, L. C. Duarte, S. Marques and R. Bogel-Lukasik, *Bioresour. Technol.*, 2010, **101**, 4775–4800.
52. G. P. van Walsum and H. Shi, *Bioresour. Technol.*, 2004, **93**, 217–226.
53. F. M. Relvas, A. R. C. Morais and R. Bogel-Lukasik, *J. Supercrit. Fluids*, 2015, **99**, 95–102.
54. F. M. Relvas, A. R. C. Morais and R. Bogel-Lukasik, *RSC Adv.*, 2015, 73935–73944.
55. T. Rogalinski, K. Liu, T. Albrecht and G. Brunner, *J. Supercrit. Fluids*, 2008, **46**, 335–341.
56. M. Sasaki, B. Kabyemela, R. Malaluan, S. Hirose, N. Takeda, T. Adschiri and K. Arai, *J. Supercrit. Fluids*, 1998, **13**, 261–268.
57. M. Sasaki, Z. Fang, Y. Fukushima, T. Adschiri and K. Arai, *Ind. Eng. Chem. Res.*, 2000, **39**, 2883–2890.
58. C. G. Liu and C. E. Wyman, *Ind. Eng. Chem. Res.*, 2003, **42**, 5409–5416.
59. T. Miyazawa and T. Funazukuri, *Biotechnol. Prog.*, 2005, **21**, 1782–1785.
60. R. C. McWilliams and G. P. van Walsum, *Appl. Biochem. Biotechnol.*, 2002, **98**, 109–121.
61. J. W. King, K. Srinivas, O. Guevara, Y. W. Lu, D. F. Zhang and Y. J. Wang, *J. Supercrit. Fluids*, 2012, **66**, 221–231.
62. M.-H. Chuang and M. Johannsen, Characterization of pH in aqueous CO_2-systems, in 9th International Symposium on Supercritical Fluids (ISSF 2009), Arcachon, France, 2009.
63. L. V. A. Gurgel, M. T. B. Pimenta and A. A. da Silva Curvelo, *Ind. Crops Product*, 2014, **57**, 141–149.
64. H. D. Zhang and S. B. Wu, *Bioresour. Technol.*, 2013, **149**, 546–550.

65. J. F. Pang, M. Y. Zheng, A. Q. Wang and T. Zhang, *Ind. Eng. Chem. Res.*, 2011, **50**, 6601–6608.
66. G. Brunner, *J. Supercrit. Fluids*, 2009, **47**, 373–381.
67. S. R. M. Moreschi, A. J. Petenate and M. A. A. Meireles, *J. Agric. Food Chem.*, 2004, **52**, 1753–1758.
68. S. R. M. Moreschi, J. C. Leal, M. E. M. Braga and M. A. A. Meireles, *Braz. J. Chem. Eng.*, 2006, **23**, 235–242.
69. R. L. Orozco, M. D. Redwood, G. A. Leeke, A. Bahari, R. C. D. Santos and L. E. Macaskie, *Int. J. Hydrogen Energy*, 2012, **37**, 6545–6553.
70. H. Yoshida, M. Terashima and Y. Takahashi, *Biotechnol. Prog.*, 1999, **15**, 1090–1094.
71. M. Firdaus and M. A. R. Meier, *Eur. Polym. J.*, 2013, **49**, 156–166.
72. X. Ning and H. Ishida, *J. Polym. Sci., Part A: Polym Chem.*, 1994, **32**, 1121–1129.
73. C. F. Wang, J. Q. Sun, X. D. Liu, A. Sudo and T. Endo, *Green Chem.*, 2012, **14**, 2799–2806.
74. D. T. Balogh, A. A. S. Curvelo and R. A. M. C. Degroote, *Holzforschung*, 1992, **46**, 343–348.
75. D. Pasquini, M. T. B. Pimenta, L. H. Ferreira and A. A. S. Curvelo, *J. Supercrit. Fluids*, 2005, **36**, 31–39.
76. A. S. R. Machado, R. M. A. Sardinha, E. G. de Azevedo and M. N. da Ponte, *J. Supercrit. Fluids*, 1994, **7**, 87–92.
77. E. Kiran and H. Balkan, *J. Supercrit. Fluids*, 1994, **7**, 75–86.
78. X. Li and E. Kiran, *Ind. Eng. Chem. Res.*, 1988, **27**, 1301–1312.
79. H. S. Lv, L. Yan, M. H. Zhang, Z. F. Geng, M. M. Ren and Y. P. Sun, *Chem. Eng. Technol.*, 2013, **36**, 1899–1906.
80. R. M. N. Roque, M. N. Baig, G. A. Leeke, S. Bowra and R. C. D. Santos, *Resour., Conserv. Recycl.*, 2012, **59**, 43–46.
81. M. M. Shah, S. K. Song, Y. Y. Lee and R. Torget, *Appl. Biochem. Biotechnol.*, 1991, **28–9**, 99–109.
82. M. H. L. Silveira, B. A. Vanelli, M. L. Corazza and L. P. Ramos, *Bioresource Technol.*, 2015, **192**, 389–396.
83. G. Cheng, P. Varanasi, C. L. Li, H. B. Liu, Y. B. Menichenko, B. A. Simmons, M. S. Kent and S. Singh, *Biomacromolecules*, 2011, **12**, 933–941.
84. V. M. Roberts, V. Stein, T. Reiner, A. Lemonidou, X. B. Li and J. A. Lercher, *Chem. - Eur. J.*, 2011, **17**, 5939–5948.
85. R. Beauchet, F. Monteil-Rivera and J. M. Lavoie, *Bioresour. Technol.*, 2012, **121**, 328–334.
86. J. R. Gasson, D. Forchheim, T. Sutter, U. Hornung, A. Kruse and T. Barth, *Ind. Eng. Chem. Res.*, 2012, **51**, 10595–10606.
87. R. W. Thring and J. Breau, *Fuel*, 1996, **75**, 795–800.
88. P. F. Britt, A. C. Buchanan, M. J. Cooney and D. R. Martineau, *J. Org. Chem.*, 2000, **65**, 1376–1389.
89. G. Dobele, I. Urbanovich, A. Volpert, V. Kampars and E. Samulis, *Bioresources*, 2007, **2**, 699–706.

90. A. Liu, Y. Park, Z. L. Huang, B. W. Wang, R. O. Ankumah and P. K. Biswas, *Energy Fuel.*, 2006, **20**, 446–454.
91. S. S. Toor, L. Rosendahl and A. Rudolf, *Energy*, 2011, **36**, 2328–2342.
92. A. M. Numan-Al-Mobin, K. Voeller, H. Bilek, E. Kozliak, A. Kubatova, D. Raynie, D. Dixon and A. Smirnova, *Energ. Fuel.*, 2016, **30**, 2137–2143.
93. M. Schrems, F. Liebner, M. Betz, M. Zeilinger, S. Bohmdorfer, T. Rosenau and A. Potthast, *J. Wood Chem. Technol.*, 2012, **32**, 225–237.
94. R. J. A. Gosselink, W. Teunissen, J. E. G. van Dam, E. de Jong, G. Gellerstedt, E. L. Scott and J. P. M. Sanders, *Bioresour. Technol.*, 2012, **106**, 173–177.
95. M. Shafiei, K. Karimi and M. J. Taherzadeh, *Bioresour. Technol.*, 2010, **101**, 4914–4918.
96. A. Barakat, H. de Vries and X. Rouau, *Bioresour. Technol.*, 2013, **134**, 362–373.
97. W.-H. Chen, B.-L. Pen, C.-T. Yu and W.-S. Hwang, *Bioresour. Technol.*, 2011, **102**, 2916–2924.
98. C. Martín, H. B. Klinke, M. Marcet, L. García, E. Hernández and A. B. Thomsen, *Holzforschung*, 2007, **61**, 483–487.
99. P. S. nee'Nigam, N. Gupta and A. Anthwal, *Biotechnology for Agro-Industrial Residues Utilisation*, Springer, 2009, pp. 13–33.
100. X. Li, T. H. Kim and N. P. Nghiem, *Bioresour. Technol.*, 2010, **101**, 5910–5916.
101. C. L. Li, B. Knierim, C. Manisseri, R. Arora, H. V. Scheller, M. Auer, K. P. Vogel, B. A. Simmons and S. Singh, *Bioresour. Technol.*, 2010, **101**, 4900–4906.
102. H. Zhang and S. Wu, *J. Chem. Technol. Biotechnol.*, 2015, **90**, 1640–1645.
103. J. S. Luterbacher, J. W. Tester and L. P. Walker, *Biotechnol. Bioeng.*, 2012, **109**, 1499–1507.
104. D. Pasquini, M. T. B. Pimenta, L. H. Ferreira and A. A. S. Curvelo, *J. Supercrit. Fluids*, 2005, **34**, 125–131.

CHAPTER 6

Processing of Lignocellulosic Biomass Derived Monomers using High-pressure CO_2 and CO_2–H_2O Mixtures

GIANLUCA GALLINA,[a] PIERDOMENICO BIASI,[b] CRISTIAN M. PIQUERAS[c] AND JUAN GARCÍA-SERNA*[a]

[a] Department of Chemical Engineering and Environmental Technology, High Pressure Processes Group, University of Valladolid, Valladolid ES-47011, Spain; [b] Process Chemistry Centre, Laboratory of Industrial Chemistry and Reaction Engineering, Åbo Akademi, Biskopsgatan 8, Turku/Åbo FI-20500, Finland; [c] Planta Piloto de Ingeniería Química, PLAPIQUI-Universidad Nacional del Sur-CONICET, Camino La Carrindanga km 7-CC 717, (8000) Bahía Blanca, Argentina
*Email: jgserna@iq.uva.es

6.1 Introduction

The biorefinery concept is analogous to the concept of an oil refinery and assumes the idea of the production of fuel, chemicals and energy from different types of biomass. The choice of biomass type to treat can be influenced by economic, environmental or geographical factors; there is also a direct dependence between the raw material, the technology used for its conversion into a usable output and the range of products that can be obtained. Lignocellulosic biomass is considered the most promising feedstock for the

Green Chemistry Series No. 48
High Pressure Technologies in Biomass Conversion
Edited by Rafał M. Łukasik

Published by the Royal Society of Chemistry, www.rsc.org

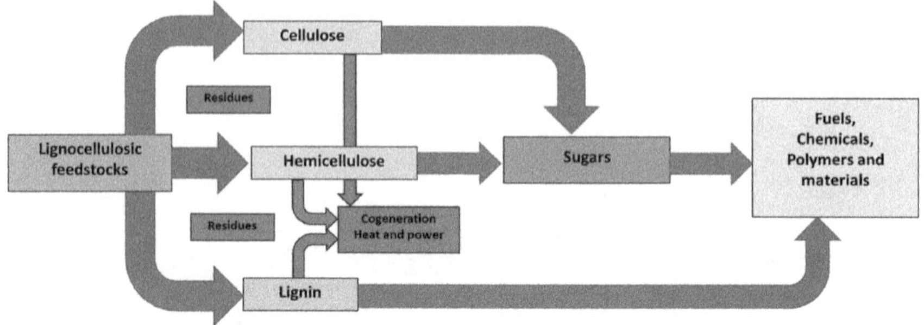

Figure 6.1 Lignocellulosic biorefinery concept outline.

production of bio-fuels and chemicals. Lignocellulosic materials belong to second-generation feedstocks, and can be obtained from various sources, such as wood residues, agricultural or municipal waste, not interfering with direct crops for human consumption. They are composed mainly of lignin, cellulose and hemicellulose, associated in a resistant structure, whose breakup requires a considerable amount of energy (moderately endothermic); however, thanks to their differentiated composition, they allow to obtain multiple products, like high value chemicals and low value but high volume fuels.[1]

Figure 6.1 shows an outline of the concept of a lignocellulosic biorefinery. Lignocellulosic feed can be pre-treated and fractionated into cellulose, hemicellulose and lignin. Lignin phenolics can be used to produce materials like plastics or adhesives, glucose from cellulose can be converted to fuels or chemicals, while other chemicals, fuels, polymers and materials can be obtained from hemicellulose. Furthermore waste cellulose, hemicellulose and lignin can be used for cogeneration.[2]

Hydrothermal pre-treatments, as well as facilitate the enzymatic attack by reducing the recalcitrance of the biomass,[3] result in the production of different compounds, through the extraction and hydrolysis of the ligno-cellulosic biopolymers.

The monomeric sugars constituting cellulose and hemicellulose, under particular conditions of temperature, acidity or residence time, can originate furfural, 5-hydroxymethylfurfural (5-HMF), glycolaldehyde, acetic acid, pyruvhaldehyde, lactic acid and other products resulting from their dehydration or through aldolic reactions.[4,5]

Although many of them are often considered undesirable by-products, as their presence inhibit a further enzymatic treatment,[6] often their commercial value exceeds that of sugars, or alcohols produced from their fermentation. For this reason, in some cases it is more correct to define these compounds as added-value products, rather than by-products. Among these chemicals, furfural, 5-hydroxymethylfurfural and lactic acid are of great economic interest: furfural can be the starting material for polymers such as nylon-6 and nylon-6,6; 5-HMF has the potential to replace terephthalic acid;[7–9] lactic acid can be used as a food preservative, flavouring agent and is also employed in pharmaceutical technology to produce water-soluble lactates.

A list of the top value added compounds from biomass was widely discussed in two volumes from NRE Laboratories,[10,11] identifying the products deriving from biomass, and the processes that would economically and technically support the production of fuels, power and chemicals in an integrated biorefinery. After a careful screening, 30 foundation chemicals were selected, which may be competitive with compounds deriving from the petrochemical industry. Some of these products, such as furfural and levulinic acid, can be directly produced *via* hydrothermal hydrolysis of biomasses, without the need for further processing. Other compounds such as formic acid, 5-HMF, acetic acid, glycolaldehyde, glyceraldehyde, pyruvaldehyde, lactic acid can be produced *via* saccharification and subsequent hydrolysis of cellulose and hemicellulose, using only water or mixtures of water and CO_2 as the only reagents, without the addition of any type of additives that presuppose a treatment and a detoxification of liquid effluents.[12,13]

In this chapter, different hydrothermal pre-treatments using subcritical and supercritical H_2O, and CO_2–H_2O mixtures will be analyzed, describing their incidence in the dehydration of monomers towards value added compounds and their further transformations. In addition, the advantages and effectiveness of using supercritical CO_2 to separate compounds as furfural and acetic acid from aqueous effluents will be explained.

Moreover different types of reactors and setups will be illustrated, describing their characteristics and their effectiveness in the hydrolysis of biopolymers and real biomasses for the production of different lignocellulosic based compounds.

6.2 Cellulose and Hemicellulose Hydrolysis

Cellulose and hemicellulose, which together with lignin are the main components of lignocellulosic biomasses, are biopolymers with different structure, and require different operating conditions to be extracted.

While hemicellulose is a branched polymer consisting in short chains of about 500 to 3000 units of different monosaccharides, cellulose presents a crystalline structure, with linear molecules composed by 7000 to 15 000 units of glucose.[14]

6.2.1 The Phenomena at a Glance

This type of reaction has a rich number of chemical engineering steps. Thus, at a molecular level one will find the bonding of the sugar monomers and aromatic monomers forming the hemicellulose, cellulose and lignin, together with the extracts, starch, essential oils, *etc.* The real reaction takes place there. At the next level, one finds the polymers of several molecular weights that can be counted using population balances. Most of the polymers will be solid, crystalline or amorphous, and the oligomers will be soluble or non-soluble depending on the number of monomers that they have got, the acetyl groups, *etc.* Next, one finds the particle level, the particles will be porous, with different shapes and sizes. They will create back-mixing due to hydrodynamics at millimetre scale. Finally, one finds the reactor level, where the particles are distributed in slurry or in a fixed bed, and heat transfer to the wall can be present, *etc.* All these levels should be considered to determine the kinetics of the reactions and to model and scale up the process.

In hot water media, the fractionation of lignocellulosic biomass takes place in solid phase, where cellulose and hemicellulose start to break into oligomers, with a decrease of their molecular weight. When a certain molecular weight is reached, they became water-soluble and the hydrolysis proceeds both in liquid and in solid phase.[15,16] Depending on the temperature and on the residence time, the oligomers extracted may undergo a fractionation process: they are depolymerized into monomers, which are subsequently decomposed in a broad range of products.

Due to the differences in structure, whereas temperatures between 140 and 190 °C are sufficient for the extraction of hemicellulose, the depolymerization of cellulose requires temperatures above 230 °C.[17–19]

6.2.2 Simple Mathematical Models to Describe Hydrolysis

There are mainly two models to study the depolymerization of the polymers: the first model considers a direct hydrolysis of cellulose or hemicellulose to the respective monomers, the second model considers the rupture of the polymers to intermediate oligomers, which are subsequently hydrolyzed to monomers;[20] since, in general, it is difficult to experimentally quantify the production of oligomers, the first model is the most extended. It assumes that the kinetics of decomposition of cellulose and hemicellulose can be summarized by two second order reactions, where k_1 represents the dissociation constant of polymers to monomers, and k_2 the reaction rate from monomers to dehydrated products.[21,22] This model, proposed by Saeman[21] for cellulose can be extended also to hemicellulose depolymerization, and does not consider any intermediate oligomer.

$$\text{polymers} \xrightarrow{k_1} \text{monomers} \xrightarrow{k_2} \text{decomposition products}$$

The concentration of monomers can be calculated analytically by integrating the two differential equations with initial value[23] $M = M_0 e^{-k_2 t} + P_0 \frac{k_1}{k_2 - k_1}(e^{-k_1 t} - e^{-k_2 t})$, where M indicates the concentration of monomers, P the concentration of polymers, t the reaction time and the index 0 the initial values. Kinetic constants can be related to the operation temperature by Arrhenius equation:[23] $k_i = k_{i0} e^{-\frac{Ea}{RT}}$, where Ea is the activation energy, R is the gas constant, T the temperature and i the kinetic coefficient (1 or 2). Typical values of k_1 and k_2 constants are represented in Table 6.1.

Table 6.1 k_1 and k_2 constants at different temperatures and pressure of CO_2.

T (°C)	k_1 (min^{-1})	k_2 (min^{-1})	p_{CO_2} (bar)	Ref.
220	0.076	0.044	0	65
235	0.116	0.052	0	65
180	0.093	0.041	0	66
180	0.127	0.064	20	66
180	0.109	0.065	35	66
180	0.073	0.068	50	66

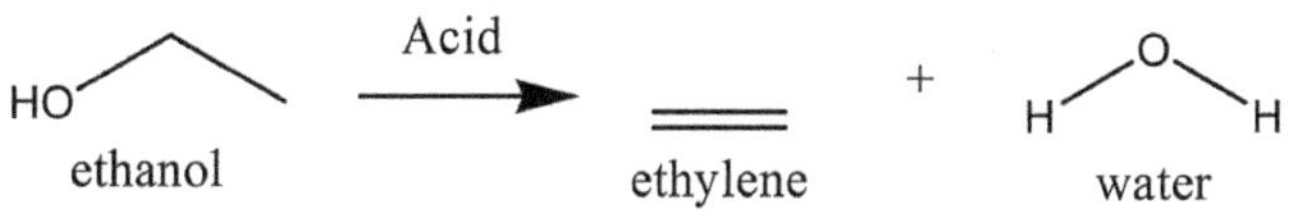

HO ethanol — Acid → ethylene + water (H–O–H)

Scheme 6.1 Example of a dehydration reaction for alcohols.

Kinetic studies carried out so far are aimed at maximising the yield of monomers, avoiding their degradation products, there are few studies that investigate deeply the formation of single degradation compounds from raw biomass;[24] there is, however, a high number of studies dealing with the formation of degradation products from pure monosaccharides.[25–27]

6.2.3 The Main Reactions of the Monomers in Water: Tautomerization, Dehydration and Aldol Reactions

The main reactions that involve the monomers of cellulose and hemicellulose in hot pressurized water are: tautomerizations, dehydrations and aldol reactions.[9]

The dehydration reaction is a chemical reaction that involves the loss of a water molecule from the reacting molecule. The hydroxyl group (–OH) is a leaving group with poor stability, a Brønsted acid often act as a catalyst, helping to protonate the hydroxyl group, producing the leaving group, $-OH_2^+$. An example of dehydration is shown in Scheme 6.1.

Tautomerization is the chemical reaction occurring when two constitutional isomers of organic compounds readily interconvert with each other. Isomers are substances with the same molecular mass and the same composition percentage of atoms, but different physical properties and often also different chemical behaviour. Many isomers have the same (or very similar) bond energy, so they easily interconvert. An example of tautomerization is shown in Scheme 6.2.

The aldol reaction is a reaction in which two molecules of an aldehyde or of a ketone, which have at least one hydrogen atom in α position to the carbonyl group (C=O), combine with each other to form a β-hydroxyaldehyde or a β-hydroxyketone. Furthermore, the product, commonly called an aldol, being very unstable, can be dehydrated and converted to the corresponding

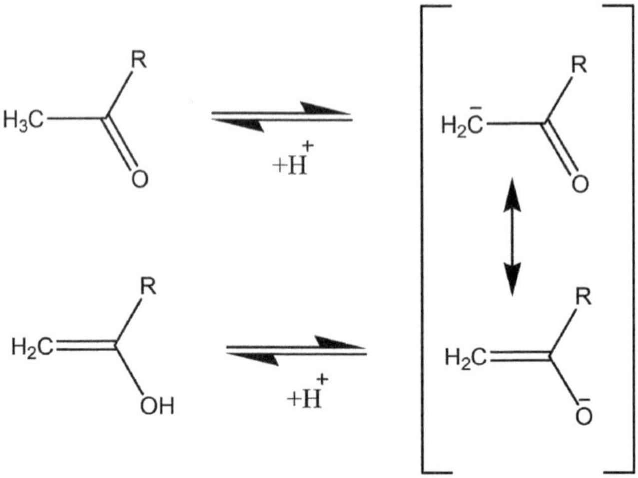

Scheme 6.2 Example of a tautomerization reaction.

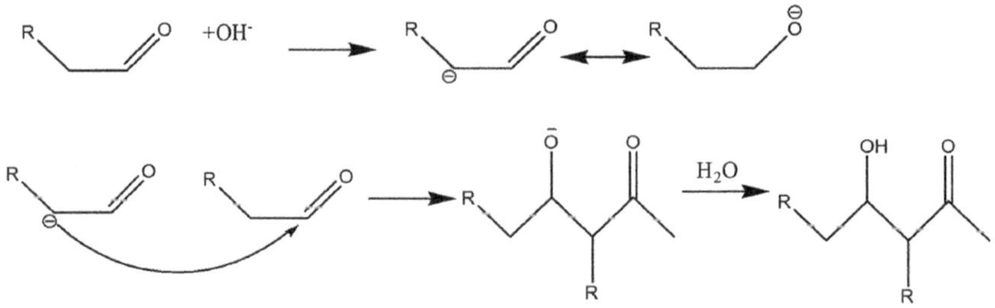

Scheme 6.3 Example of an aldol reaction.

unsaturated conjugate compound. Scheme 6.3 shows an example of aldol reaction.

The hydrolysis of hemicellulose in water begins at temperatures around 100 °C, the two principal reactions are reported in Scheme 6.4 for xylans:[28]

- Acetyl groups bonded to oligomers are cleaved and reduced to acetic acid, catalyzing the autohydrolysis of the oligomers;
- The oligomer chain is hydrolyzed to obtain monomeric sugars, which are dehydrated to furfural at harsher conditions.

As stated above, for temperatures above 100 °C, hydrolysis of hemicellulose begins and can be more or less intense, depending on the composition of the treated biomass. Xylans from hardwood contain greater amounts of acetyl groups respect to softwoods and herbaceous plants. Approximately 60% of the units of xylose, carry an acetyl group attached to position C-2 or C-3.[29] For this reason, auto-hydrolytic capacity in hemicellulose from hardwood is much more intense compared to that of other species. The dehydration of xylose to furfural begins around 160 °C, and gradually

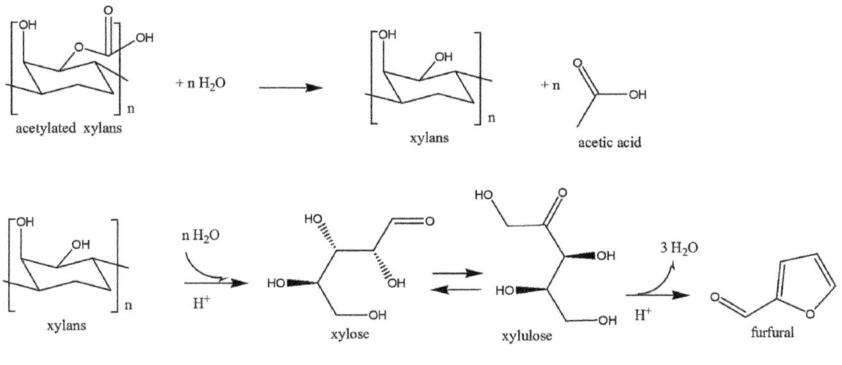

Scheme 6.4 Hydrolysis of xylans and dehydration to furfural.

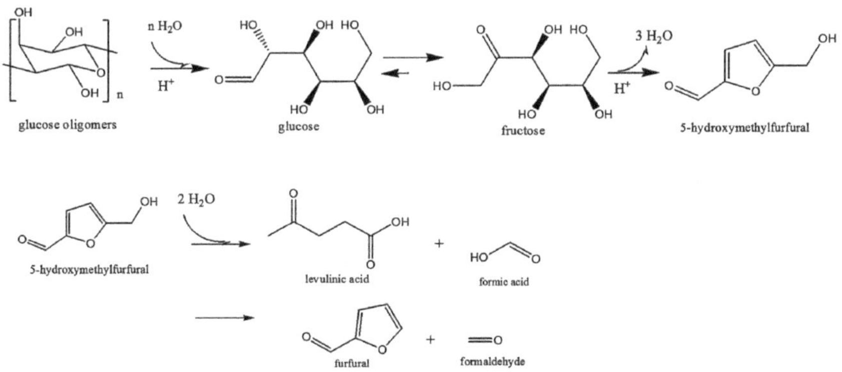

Scheme 6.5 Hydrolysis of glucose oligomers, formation and decomposition of 5-HMF.

becomes more pronounced as the temperature of the water increases.[30] Kinetic studies show that the xylose, before the dehydration to furfural, passes through an isomerization process, forming xylulose.[31,32]

As mentioned previously, cellulose structure presents a crystalline domain; strong intra-chain hydrogen bonds result in a stable and linear configuration of the fibrils[20] that require high temperatures for the depolymerization down to glucose units. Once formed, among the glucose molecules, reactions of isomerization and dehydration take place at temperatures above 200 °C.[33]

Glucose undergoes isomerization to fructose, the reverse reaction (fructose to glucose) is almost inhibited. Fructose dehydration leads to the formation of 5-hydroxymethylfurfural (5-HMF) and its formation is directly proportional to the temperature rise of the medium until about 350 °C. 5-HMF can follow principally two pathways: it can be decomposed into furfural and formaldehyde, or can be hydrated to form levulinic acid and formic acid. Reaction pathways are proposed in Scheme 6.5.[9]

In addition to the reactions listed above, the glucose molecule is also subject to aldol reactions, producing one molecule of two carbons and one molecule of four carbons; fructose produces two molecules of three carbons (like glyceraldehyde).[34]

Glucose, through an aldol reaction, can lead to the production of erythrose and glycolaldehyde. Fructose, through a first aldol reaction can produce glyceraldehyde and its isomer dihydroxyacetone. Another aldol reaction and subsequent dehydration can convert the glyceraldehyde into pyruvaldehyde, which can be further converted into lactic acid, with the loss of another water molecule. Dehydration as well as isomerization reactions are favoured in acidic media,[34] while aldol reactions are promoted in neutral media.[35] Principal aldol reactions involving glucose and fructose monomers are represented in Scheme 6.6.

Also xylose can participate in aldol reactions inside a hydrothermal medium, producing a molecule with three atoms of carbon and a molecule with two atoms of carbon (Scheme 6.7). Glyceraldehyde, consequently,

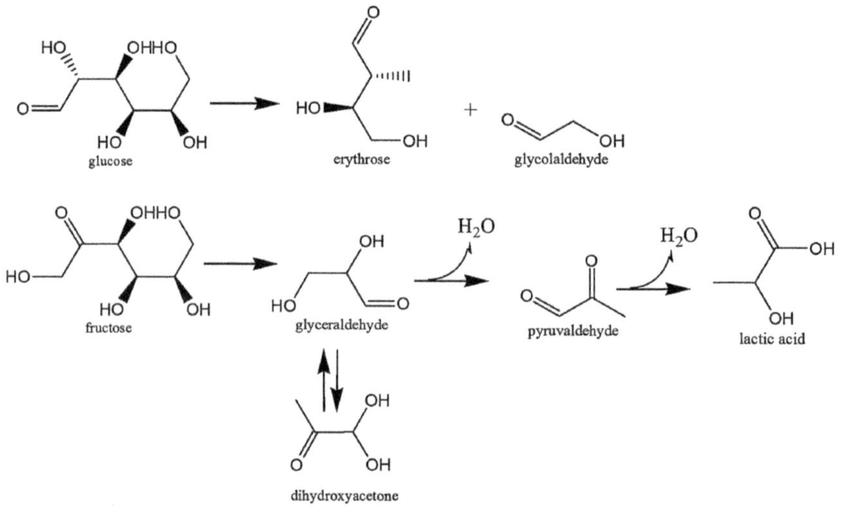

Scheme 6.6 Aldol reactions involving glucose and fructose.

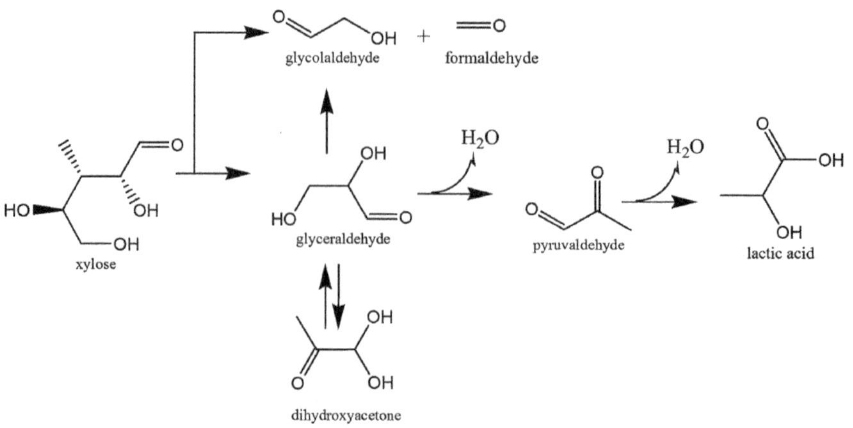

Scheme 6.7 Aldol reactions involving xylose.

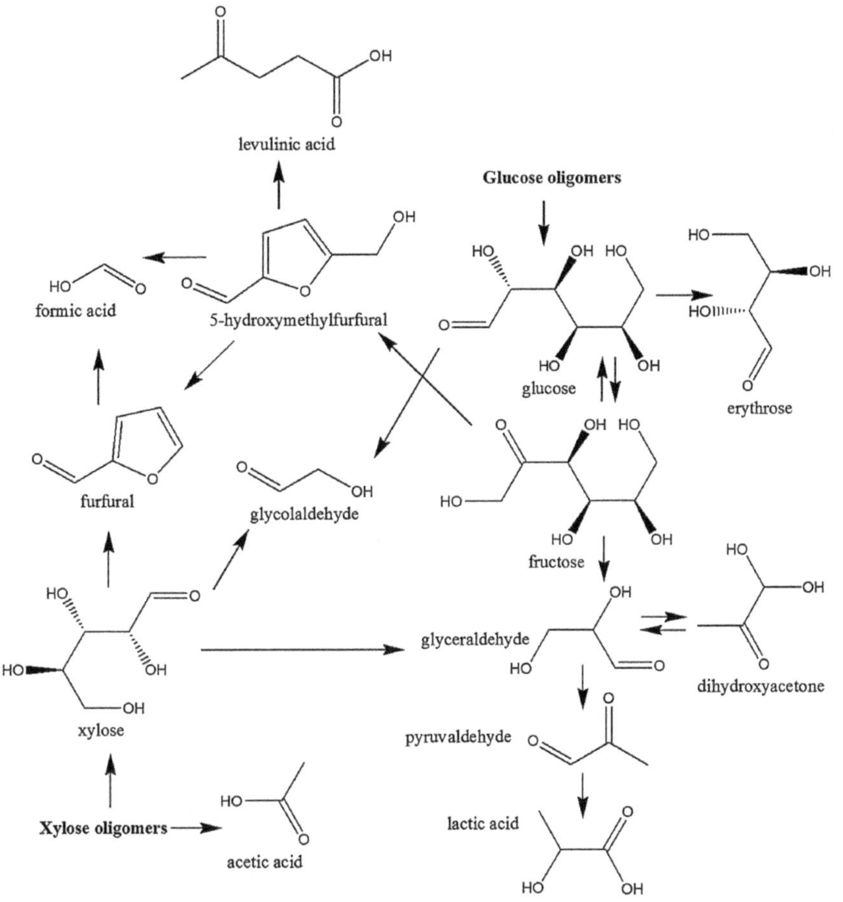

Scheme 6.8 Reaction pathway for hydrolysis of xylose and glucose oligomers.

through another aldolic reaction can lead to the production of glycolaldehyde and formaldehyde, or can be isomerized into dihydroxyacetone.

Scheme 6.8 shows schematically a simplified pathway for compounds resulting from the depolymerization of cellulose and hemicellulose oligomers and decomposition of the monomers (xylose oligomers were chosen as representatives of hemicellulose oligomers).[5,13,26,32,36–38]

6.3 Reaction Medium and Operational Conditions

As mentioned, there are not many studies with the main objective to carry out pre-treatments for lignocellulosic biomass to maximize the conversion of carbohydrate to form dehydration products, since often the main purpose is to avoid the so-called "degradation"; however, there are conditions that promote the formation of these chemicals.

What are the main factors influencing the reactions of depolymerization and hydrolysis of lignocellulosic biomass in aqueous media?

- Temperature
- Residence/reaction time of the compounds inside the reaction medium
- pH at which the reaction takes place (*i.e.* proton concentration)

The different properties that the pressurized water may assume varying these conditions, can be used to control the selectivity of the reactions, in other words, to tune the reaction. According to the idea of thermohydrolysis reactions performed using pure water, without addition of mineral acids or other chemical additives that presuppose a subsequent disposal or neutralization of the sludge, two kind of aqueous media can be considered: subcritical water (high polarity, medium temperature) and supercritical water (low polarity, high to very high temperature). By choosing the appropriate temperature and pressure it is therefore possible to obtain a solvent with distinct and peculiar characteristics, which can promote some reactions than other. The addition of carbon dioxide to the aqueous medium allows more possibilities to influence the kinetics of the reactions, by increasing the conversion of oligosaccharides and modifying the selectivity towards certain products.

In this section we will study the aqueous reaction media mentioned above; we will see which conditions to adopt, in order to achieve high selectivity towards value-added products from lignocellulosic biomasses.

6.3.1 Subcritical Water and Carbonated Subcritical Water

It was verified that the addition of CO_2 in hot pressurized water enhances the hydrolysis of lignocellulosic materials[39,40] due to the decrease in pH, by the formation and the dissociation of carbonic acid in the aqueous mixture.[22] When formation of carbonic acid occurs, an increase in hydronium ion concentration is observed due to the dissociation of the unstable acid, thus promoting acid-catalyzed dissolution of the biomass[41] and allowing reduction of the temperature and the reaction time of the reactions.

By controlling the amount of CO_2 dissolved in water, which is dependent on the temperature and the pressure of the mixture, the pH of the solution mixture can be controlled. Over the temperature range 25 to 70 °C and pressure range 70 to 200 bar, pH may decrease to values down to 2.8.[42]

Figure 6.2 shows the variation of solubility of CO_2 in water at different temperatures and pressure according to experimental data.[43]

Unlike the pre-treatments with acids, the addition of CO_2 does not require a subsequent neutralization processes, because the CO_2 removal can be accomplished by the reduction in pressure and CO_2 desorption.

In an aqueous pre-treatment, the combination of temperature and reaction time define the severity of the reaction,[44] according to the experimental

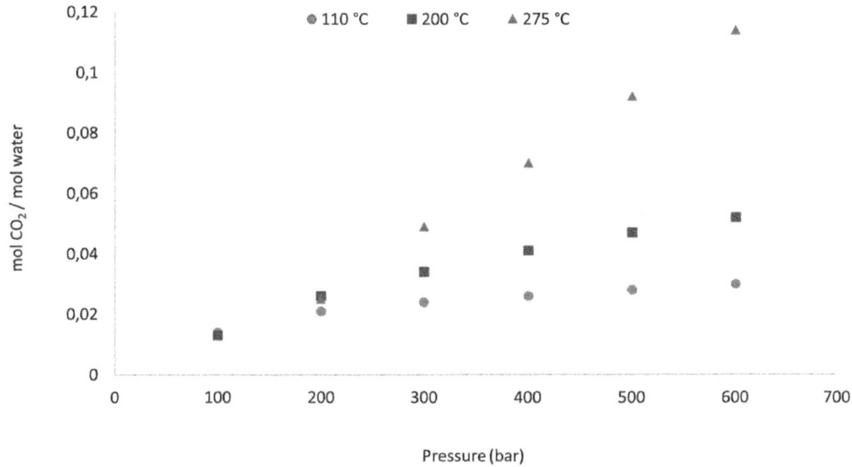

Figure 6.2 Solubility of CO_2 in water at different temperatures and pressures. Data taken for ref. 43.

equation[45] $R_0 = t * \exp[(T - 100)]/14.75]$, where t is the reaction time (min), and T is the temperature (°C) (Figure 6.3).

If carbon dioxide is added to water, the pH variation changes the severity of the reaction; it is therefore necessary to modify the factor taking into account of this influence. Van Walsum *et al.* defined a combined severity factor for binary systems with water and CO_2 in the temperature range of 100–250 °C and partial pressure of CO_2 up to 151.9 bar, suggesting the equation:[46] $CS_{P_{CO_2}} = \log(R_0) - 8.00 * 10^{-6} * T^2 + 0.00209 * T - 0.216 * \ln(p_{CO_2}) + 3.92$, in which $CS_{P_{CO_2}}$ is the severity due to the presence of CO_2, R_0 is the severity factor calculated as above, p_{CO_2} is the partial pressure of CO_2 (atm) and T is the temperature (°C). This combined equation includes all the most important parameters affecting the hydrolysis of oligomers to monomers, and to dehydration products.

Some of the early experiments in which carbon dioxide was added to the aqueous medium to improve the efficiency of hydrothermal pre-treatments were carried out by Van Walsum *et al.* As well as studying the production of monosaccharaides, they investigated the formation of degradation products in lignocellulosic pre-treatments with mixtures of water and CO_2.[47] They analyzed the composition of the extraction liquors from experiments in a batch reactor, at 180 °C, after a cooking time of 16 min; comparing the data obtained using only water as a solvent and water with 55 bar of CO_2. Two biomasses were tested: corn stover and aspen wood, Figure 6.4 shows the differences in concentration between the experiments without CO_2 and with CO_2 for both the biomasses and for the different compounds.

The two biomasses had a different behaviour: while in aspen wood the addition of CO_2 resulted in an increase of all the products analyzed (despite the weak analytical reproducibility), in corn stover it led to the increase in

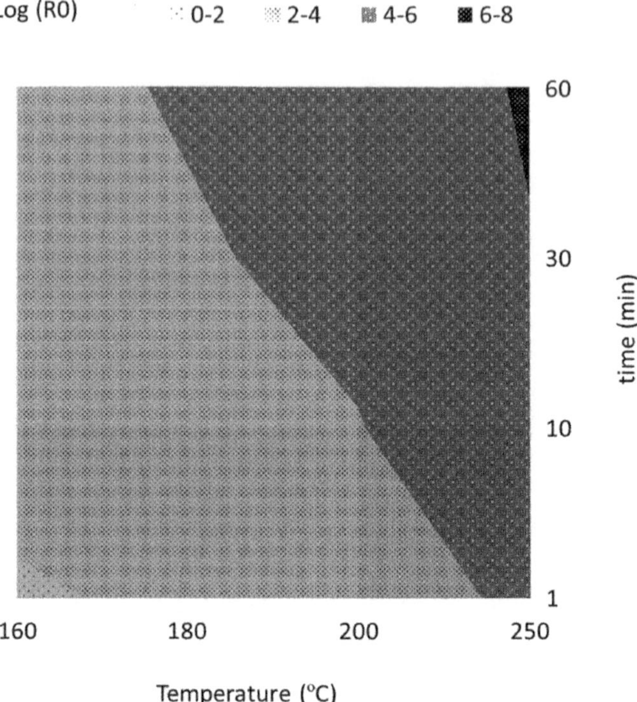

Figure 6.3 Values of $\log(R_O)$ varying temperature and residence time.

the concentration of some products and to the decrease of others. Acetic acid concentration increased in both biomasses, confirming the capacity of CO_2 to hydrolyze xylan oligomers, even if in the case of aspen wood the yield of xylose monomers did not seem to show relevant differences if compared with experiments with only water.[48] In corn stover, furfural and formic acid concentrations decreased when adding CO_2 while in aspen wood seemed to increase. This behaviour should indicate that in aspen wood, the autohydrolytic capacity of the raw material in water was sufficient to break the hemicellulose oligomers into xylose, and the acidification of the aqueous medium due to the presence of CO_2 leads to the degradation of the monomers.[48] In the case of corn stover, the autocatalytic capacity was less strong, and CO_2, with its capacity of penetrating small pores of recalcitrant lignocellulosic structure, favoured the extraction of hemicellulose oligomers from the wooden matrix and with its acidifying effect, led to the hydrolysis of the oligomers into monosaccharaides, more than the degradation of the lasts.[39] The hydrolyzing power of carbonic acid, therefore, seemed to exert a different effect depending on the type of pre-treated biomass.

Aside from the considerations made so far, the hydrolyzing effect due to CO_2 in aqueous media, appears to depend very much on the partial pressure of the gas, the temperature of the system and thus the solubility of CO_2 in water. Figure 6.2 indicates that the solubility of CO_2 in water increases with

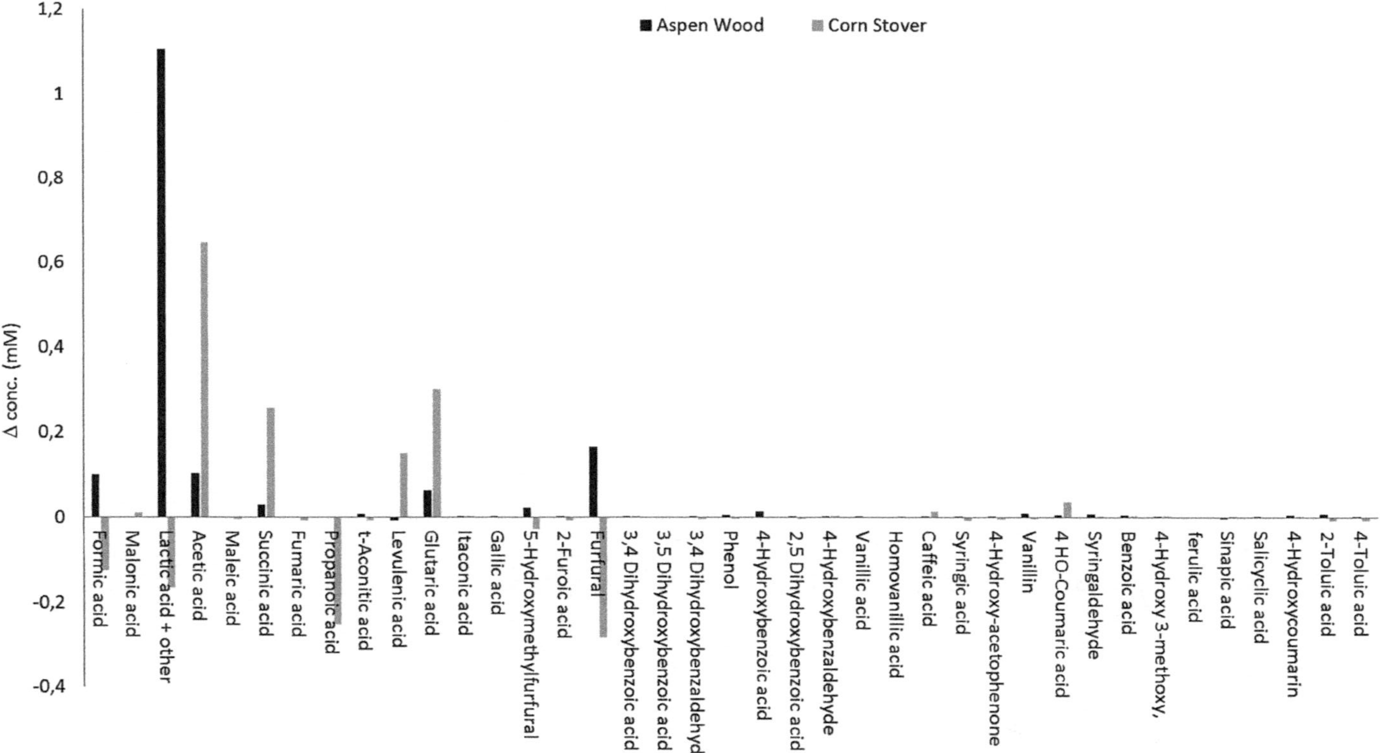

Figure 6.4 Concentration differences between experiments with water–CO_2 mixtures and only water. Data obtained from ref. 47.

increasing pressure, in addition, with the operating conditions tested in the experiments, the solubility of CO_2 increases at increasing the temperature. To obtain a catalytic effect due to the presence of carbon dioxide in the hydrolysis of lignocellulosic compounds, it is therefore necessary to operate with sufficiently high pressures.

In the hydrothermal pre-treatment of wheat straw with a temperature of 210 °C, Magalhães da Silva *et al.*, found an increase in the production of furfural from a concentration of 0.1 g L^{-1} to 5.4 g L^{-1} when adding CO_2 at 60 bars to the aqueous medium;[49,50] with the same raw material, increasing the temperature at 225 °C, the same group noticed only a small difference between the amount of furfural when increasing the initial pressure of CO_2 (Figure 6.5).[46,51]

Of great interest are the experimental results obtained by Rogalinski *et al.*[52] and King *et al.*[53] The two groups studied the hydrolysis of similar biomass (rye straw and switchgrass, respectively) in mixtures of water under subcritical conditions with the addition of carbon dioxide at different pressures. Pressure of carbon dioxide was 0 and 100 bar in the experiments of the first group, whereas ranged from 150 to 550 bar in the experiments of the second group. While no significant differences were reflected between experiments performed with pure water and those made with carbonated water at 100 bar, an increased catalytic effect was recognized with pressure: the yield of xylose increased approximately 2 wt%, by increasing the pressure from 350 bar to 450 bar at constant temperature of 170 °C, a value only slightly smaller than that obtained using dilute sulfuric acid hydrolysis. Moreover, at these conditions, the hydrolysis of switchgrass resulted in a production of a larger amount of 5-HMF and furfural respect to those produced using dilute acid pre-treatment. An enhance in the hydrolysis of cellulose by means of carbon dioxide under high pressure was also confirmed by other experiments carried out by the group of Brunner.[54]

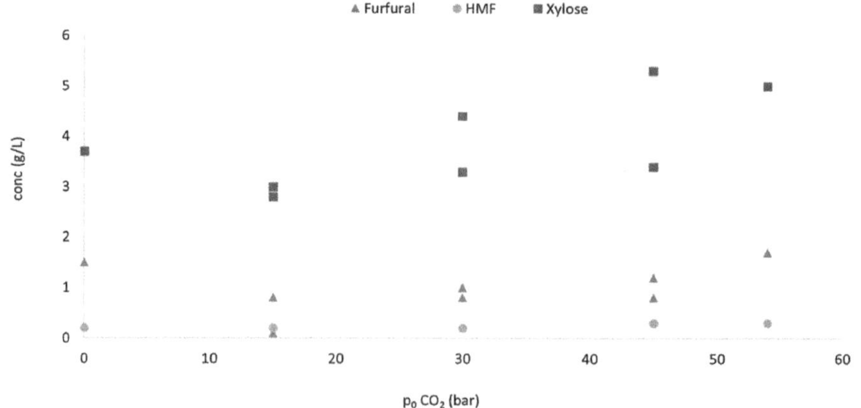

Figure 6.5 Concentration of furfural and HMF in water mixtures with different pressures of CO_2.
Data obtained from Morais *et al.*[51]

The addition of carbon dioxide to water, if carried out at sufficiently high pressures, seems to have a catalytic effect in the hydrolysis of oligosaccharides and in the decomposition of monosacharaides; it is necessary to analyze which is the combined effect of the temperature change and the carbonation, and which prevails over the other.

Dhamdere *et al.*[55] tested the hydrolysis of switchgrass at different temperatures (220, 250, 280, 310 °C) in a semi-continuous batch flow system, with and without the addition of CO_2 with a pressure of 68 bars. Results indicated that at temperatures between 220 °C and 280 °C, the production of furfural was higher in the water–CO_2 mixture than in pure water; when the temperature was increased to 310 °C the difference in the yield between the two media was only minimal, indicating that at that temperature there was no catalytic activity due to the carbonated water.[55] A maximum yield of 1.3 wt% was obtained at 310 °C.

At all the temperatures tested, the production of 5-HMF was enhanced in carbonated water mixtures, as 5-HMF is a product deriving from the dehydration of glucose, and thus from cellulose, its production is favoured at high temperatures. Yield of 5-HMF was similar at 280 °C and at 310 °C, indicating that most of the formation occurs at 280 °C. A maximum yield of 1.8 wt% was obtained at optimum conditions.

The main operating condition that controls the hydrolysis of lignocellulosic biomass is therefore the temperature. Hemicellulose and cellulose polymers are extracted from biomass and fractionated into oligomers for effect of temperature, carbon dioxide is proven to have a slight additional role in this phase, by swelling the plant material, and favouring the breakage.[56] The rupture of the oligomers is always influenced for the most part by the temperature, however, the addition of carbon dioxide with sufficiently high pressures, can play a significant supporting role.

Temperature is also the main responsible for the decomposition of the monomers: the increase of temperature leads to a higher conversion of xylose and glucose to furfural and 5-HMF, and their subsequent processing in the other products depicted in Scheme 6.8. The addition of carbon dioxide has a catalytic effect in the production of furfural at temperatures below 300 °C, at higher temperatures its incidence is negligible. The dehydration of glucose to HMF is catalyzed by carbon dioxide up to temperatures above 310 °C.

Summarizing what has been said so far, carbon dioxide at high pressures can be used in conjunction with pressurized hot water to enhance the hydrolysis of lignocellulosic oligomers to obtain monomers, and to catalyze the dehydration of monosaccharaides to obtain added value products.

Experiments conducted starting from pure xylose as a raw material, indicate that a conversion of 97% is achieved in water at 230 °C with CO_2 at 12 MPa; with a yield of 68% to production of furfural. Lower pressures lead to lower yields of furfural, while higher pressures did not show significant changes.[25] The initial concentration of xylose in the reaction influences the production of furfural; optimal concentrations of monomer to obtain a high selectivity, are around 4%, higher concentrations lead to lower yields.

Carbon dioxide, in addition to favour the production of monosaccharaides dehydration compounds, can also facilitate the removal from the aqueous phase and recovery, in particular that of furfural. The use of supercritical CO_2 for extracting low concentration of furfural (around 1 wt%) from aqueous solutions, is a good alternative to organic solvents.[57] A 48.1% of produced furfural can be extracted with CO_2 at 8 MPa from an aqueous mixture with a temperature of 230 °C; higher temperatures decrease the extraction solubility, as furfural solubility in CO_2 decreases with increasing the temperature. The increase in the concentration of xylose in the reactor is proportional to the concentration of furfural extracted; depending on whether the goal is to recover furfural or increase the yield, different concentrations of initial xylose can be selected.[25,57]

6.3.2 Reactions in Supercritical Water

Water is under supercritical conditions at temperatures higher than 374 °C and pressures higher than 22.1 MPa. The intermolecular structure of water, at these conditions, varies significantly as hydrogen bonds are significantly reduced in number, giving both gas-like properties (like high diffusivity and low viscosity) and liquid-like properties like high density.

The dielectric constant is subjected to significant variations, reaching at supercritical conditions values inside the common range of most organic solvents. At normal conditions of 25 °C and pressure of 1 bar, water has a dielectric constant of about 78.[58] Figure 6.6 shows that at a pressure of 25 MPa bar and temperature of 375 °C the dielectric constant of water is around 12, and decreases rapidly at increasing the temperature. A range

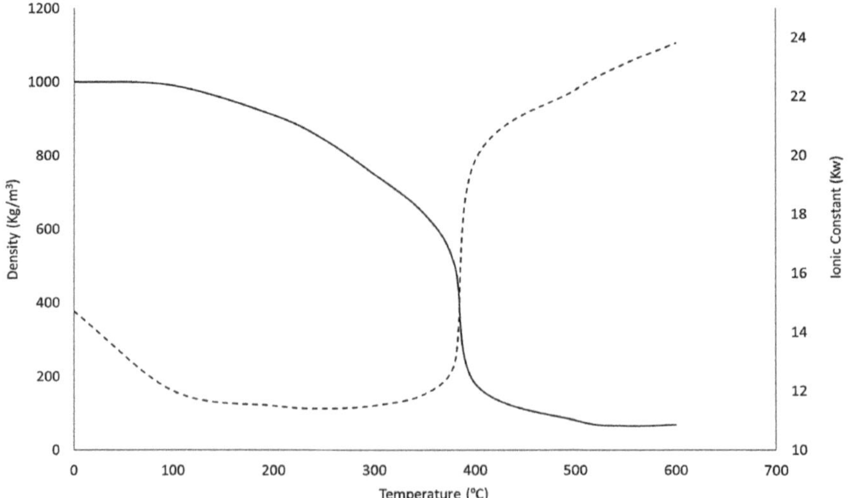

Figure 6.6 Properties of water at subcritical and supercritical at 25 MPa. Data taken from ref. 60, 63, and 64.

between 2 and 30 is typical for most organic solvents for dissolving organic macromolecules such as cellulose.

Another property which varies significantly at supercritical conditions is the ionic product of water (K_w). At ambient conditions, the value of K_w, represented as the product of H^+ and OH^- concentrations, is 10^{-13}, at temperatures around 300 °C it reaches its maximum value (10^{-11}), which creates a medium with high ions concentration, favouring acid–base catalyzed reactions. Under supercritical conditions, K_w decreases drastically (to 10^{-25}),[59] promoting non-ionic reactions.[58,60]

The combination of these properties and the high temperature that allows high reaction rates makes supercritical water an effective solvent and an excellent reaction medium in the hydrolysis processes of lignocellulosic materials. At near critical conditions, therefore, the drastic changes in the ionic product and in the density of water influences the reaction of degradation of monosaccharaides. In the case of xylose, while at subcritical temperatures, the main product of degradation is furfural, with a maximum production between 300 and 350 °C, at supercritical conditions aldolic reactions are favoured, with an increase in the formation of glycolaldehyde and glyceraldehyde. A maximum yield of retro-aldol products of around 45 wt% obtained at 400 °C and 100 MPa.[31,32]

Similarly, with glucose as base monomer, the dehydration reaction leading to the formation of 5-HMF is promoted at subcritical conditions with a maximum at around 350 °C, while retro-aldol condensations are favoured at higher temperatures, with water under supercritical conditions; glycolaldehyde is the main product deriving from glucose, and at 450 °C and 35 MPa, yields as high as 70 wt% can be reached.[61]

Also the pressure can influence the dehydration reactions, as low densities favour aldolic reactions, while high densities and high ionic products enhance the formation of furfural and 5-HMF. As stated by Aida *et al.*, higher yields of 5-HMF from glucose (8 wt%) were obtained at 350 °C and 80 MPa, compared to other experiments performed at the same temperature and lower pressures; also the furfural yield reaches the maximum value at the maximum pressure tested, of 80 MPa.[9,27] Pressures higher than 100 bar, under supercritical conditions (400 °C) promote also the formation of lactic acid, showing a decrease of the yield of glyceraldehyde, dihydroxyacetone and pyruvaldehyde.

Even if 5-HMF and furfural achieve the maximum production at subcritical conditions, the use of supercritical water as reaction medium allows the intensification of the process by reducing the required residence time of the reactions. Typical residence times of glucose and fructose reactions in water at near critical and supercritical conditions are between 20 ms and 5 s. These residence times are three orders of magnitude lower than those required at low temperature catalyzed processes and up to five orders of magnitude lower than those needed in microorganism processes.[35] At 400 °C and 100 MPa, glycolaldehyde can be produced from xylose with residence times of 0.5 s with a yield of 40%, and; lactic acid from glucose at 400 °C requires residence times between 10 s and 20 s to be produced,[35] while at reaction temperatures around 300 °C will require residence times of 60 s.[62]

6.4 Reaction Configuration

What is the best reaction technology for each case study? When facing the question of what type of equipment will be the best option to study a subcritical or supercritical biomass fractionation or hydrolysis there are several guidelines that might help.

1. Subcritical or supercritical? How much time do I need?

   The products from the hydrolysis will highly depend on the residence time that is used, provided the residence time will depend on temperature too. If we are using subcritical water (*e.g.* typically between 160 °C and 250 °C) hemicellulose usually takes from 10 to 20 min to hydrolyze, while cellulose takes from 40 to 60 min at least. In that case the use of a batch reactor is feasible, although heating-up takes 10 to 20 min and cooling down might take 5 min, and temperature-time profile must be considered in kinetics calculation. The use of a semi-continuous reactor is also valid as the real residence time of the liquid phase is usually less than 1 min, although the solid rests in the reactor until it is opened.

   When supercritical water is used the temperature goes over 374 °C, but even operating at 300 °C (not supercritical) accelerates the hydrolysis so much that the produced sugars degrade quite easily and sometimes even re-polymerize. In such cases, using batch-wise is not recommended. Semi-continuous at high flowrate is an option, but some degradation might appear. The perfect system will be the continuous reactor, but for this, milling under 200–300 um is usually needed, as a slurry needs to be pumped at high pressure. To assure low residence times (*e.g.* 20 to 40 ms, less than 1 s always) an effective system for heating-up and quenching is required.
2. How much sample do I need to analyze? How much power do I need?

   Analyzing the products from biomass hydrolysis is expensive and time consuming, as it normally requires further hydrolysis, *e.g.* using NREL procedures. The question is, is my system reliable? Biomass exhibit great differences and it is important to be sure having a representative for the experiment. Common subcritical batch reactors are between 25 mL and 1 L (not many cases of 5 or 20 L), supercritical batch are normally below 250 mL. You can considered to fill the reactor 50–60% with water and a water/biomass mass ratio of 20 : 1, this means treating 6.25 g of biomass in a typical 250 mL (*e.g.* Autoclave, Parr, *etc.*). For that case, you should expect 40–50% sugars yield, so a total of 2 to 3 grams of sugars (pentoses and hexoses).

For the continuous reactor the mass of solid obtained depends on the flowrate and the time for sample collection. It is typical a 10% solid concentration at the inlet. Regarding the flowrate consider your electric power, as you will need approx. 65 W per mL min^{-1} of flowrate (*e.g.* 20 mL min^{-1} of liquid inlet will require 1.3 kW). This will be acute for a supercritical continuous reactor that requires high flowrates to reduce the residence times.

Semi-continuous reactors are similar to the batch-wise in terms of solid handled and similar to the continuous in terms of power needed.

3. Do I need to mill?

   Batch and semi-continuous do not require extensive milling, if you need to keep the particle as it is, *e.g.* beans, chips, pellets, *etc.* batch and semi-continuous can be the option. You will only need a filter to hold the particle bed.

   On the other hand, a continuous operation requires very fine particles. Laboratory scale equipment (*e.g.* pipes of 1/8″, 1/4″, *etc.*) pumps like HPLC pumps, or membrane pumps can handle some solids (not always, check with your provider) under 100 microns. The bottleneck is normally the checkvalves of the pump. At a higher scale, *i.e.* pilot or demo, there are pumps that can even handle 1 mm particles in slurry.

4. Can or should I recycle the CO_2?

If you have used CO_2 to enhance the reactivity and hydrolysis, as explained in the book, you will probably need to recover it to improve your economic balance. At laboratory scale recovering is probably not recommended and not needed, but at pilot scale (*e.g.* a 5 L reactor) it starts being an issue. It will require basically a high-pressure vapour-liquid separator to split the two phases and a condenser to recover CO_2 as a liquid for easier pumping. There are many examples of these options in the reference books for supercritical fluid extraction, like the one from G. Brunner.[67]

6.5 Conclusions

In this chapter it has been shown that temperature is the main operating variable to manipulate for modifying the kinetics of the reactions that lead to the rupture of oligosaccharides, and to the formation of degradation products in aqueous media. The addition of carbon dioxide at high pressures, in water at subcritical conditions, enhances the hydrolysis of lignocellulosic compounds, through the formation of carbonic acid and its subsequent dissociation. The decompression of the carbon dioxide at the end of the extraction process, moreover, facilitates the removal and recovery of compounds such as furfural from the aqueous mixture.

The pre-treatment with mixtures of water and CO_2 constitute therefore an effective way to produce value added products from lignocellulosic biomass.

References

1. D. King, O. Inderwildi, A. Williams and A. Hagan, *The future of industrial biorefineries, World Economic Forum*, Geneva, Switzerland, 2010.
2. B. Kamm, P. R. Gruber and M. Kamm, *Biorefineries-industrial processes and products. Status Quo and Future Directions*, Wiley-VCH Verlag GmbH & Co. KGaA, Weinheim, 2006.
3. Y. Pu, F. Hu, F. Huang, B. Davison and A. Ragauskas, *Biotechnol. Biofuels*, 2013, **6**, 15.
4. F. Cherubini, *Energ. Convers. Manage.*, 2010, **51**, 1412–1421.
5. P. Mäki-Arvela, B. Holmbom, T. Salmi and D. Y. Murzin, *Catal. Rev.*, 2007, **49**, 197–340.
6. M. S. Moreno, F. E. Andersen and M. S. Díaz, *Ind. Eng. Chem. Res.*, 2013, **52**, 4146–4160.
7. F. W. Lichtenthaler and S. Peters, *C. R. Chim.*, 2004, **7**, 65–90.
8. C. Moreau, M. N. Belgacem and A. Gandini, *Top. Catal.*, 2004, **27**, 11–30.
9. T. M. Aida, Y. Sato, M. Watanabe, K. Tajima, T. Nonaka, H. Hattori and K. Arai, *J. Supercrit. Fluid.*, 2007, **40**, 381–388.
10. T. Werpy and G. R. Petersen, *Top Value Added Chemicals From Biomass – Volume I: Results of Screening for Potential Candidates from Sugars and Synthesis Gas* the Pacific Northwest National Laboratory (PNNL) and the National Renewable Energy Laboratory (NREL), http://www.nrel.gov/docs/fy04osti/35523.pdf, 2004.
11. J. E. Holladay, J. J. Bozell, J. F. White and D. Johnson, *DOE Report PNNL-16983 (website: http://chembioprocess.pnl.gov/staff/staff_info.asp)*, 2007.
12. P. Mäki-Arvela, T. Salmi, B. Holmbom, S. Willför and D. Y. Murzin, *Chem. Rev.*, 2011, 111, 5638–5666.
13. D. A. Cantero, M. D. Bermejo and M. J. Cocero, *J. Supercrit. Fluids*, 2013, **75**, 48–57.
14. L. J. Gibson, *J. R. Soc. Interface*, 2012, 1–18.
15. A. Cabeza, C. Piqueras, F. Sobrón and J. García-Serna, *Bioresour. Technol.*, 2016, **200**, 90–102.
16. G. Gallina, Á. Cabeza, P. Biasi and J. García-Serna, *Fuel Process. Technol.*, 2016, **148**, 350–360.
17. D. A. Cantero, M. D. Bermejo and M. J. Cocero, *J. Supercrit. Fluids*, 2015, **96**, 21–35.
18. A. Romaní, G. Garrote, F. López and J. C. Parajó, *Bioresour. Technol.*, 2011, **102**, 5896–5904.
19. O. Bobleter, *Prog. Polym. Sci.*, 1994, **19**, 797–841.
20. L. Negahdar, I. Delidovich and R. Palkovits, *Appl. Catal., B*, 2016, **184**, 285–298.
21. J. F. Saeman, *Ind. Eng. Chem.*, 1945, **37**, 43–52.
22. C. Schacht, C. Zetzl and G. Brunner, *J. Supercrit. Fluids*, 2008, **46**, 299–321.
23. R. Aguilar, J. Ramırez, G. Garrote and M. Vázquez, *J. Food Eng.*, 2002, **55**, 309–318.

24. S. Rivas, M. J. González-Muñoz, V. Santos and J. C. Parajó, *Bioresour. Technol.*, 2014, **162**, 192–199.
25. K. Gairola and I. Smirnova, *Bioresour. Technol.*, 2012, **123**, 592–598.
26. B. M. Kabyemela, T. Adschiri, R. M. Malaluan and K. Arai, *Ind. Eng. Chem. Res.*, 1999, **38**, 2888–2895.
27. T. M. Aida, K. Tajima, M. Watanabe, Y. Saito, K. Kuroda, T. Nonaka, H. Hattori, R. L. Smith and K. Arai, *J. Supercrit. Fluids*, 2007, **42**, 110–119.
28. G. Garrote, H. Dominguez and J. C. Parajo, *J. Chem. Technol. Biotechnol.*, 1999, **74**, 1101–1109.
29. A. Teleman, M. Nordström, M. Tenkanen, A. Jacobs and O. Dahlman, *Carbohydr. Res.*, 2003, **338**, 525–534.
30. S. Noguki, R. Uehara, M. Sasaki and M. Goto, Thermal stability of monosaccharides in subcritical water, in 8th International Symposium on Supercritical Fluids Kyoto, Japan, 2006.
31. N. Paksung and Y. Matsumura, *Ind. Eng. Chem. Res.*, 2015, **54**, 7604–7613.
32. T. M. Aida, N. Shiraishi, M. Kubo, M. Watanabe and R. L. Smith, *J. Supercrit. Fluids*, 2010, **55**, 208–216.
33. T. Zhang, Glucose production from cellulose in subcritical and supercritical water, University of Iowa, 2008.
34. R. S. Assary, T. Kim, J. J. Low, J. Greeley and L. A. Curtiss, *Phys. Chem. Chem. Phys.*, 2012, **14**, 16603–16611.
35. D. A. Cantero, Intensification of cellulose hydrolysis process by supercritical water: Obtaining of added value products, University of Valladolid, 2014.
36. B. M. Kabyemela, T. Adschiri, R. Malaluan and K. Arai, *Ind. Eng. Chem. Res.*, 1997, **36**, 2025–2030.
37. N. Akiya and P. E. Savage, *Chem. Rev.*, 2002, **102**, 2725–2750.
38. H. Rasmussen, H. R. Sorensen and A. S. Meyer, *Carbohydr. Res.*, 2014, **385**, 45–57.
39. G. P. van Walsum and H. Shi, *Bioresour. Technol.*, 2004, **93**, 217–226.
40. J. S. Luterbacher, Q. Chew, Y. Li, J. W. Tester and L. P. Walker, *Energ. Environ. Sci.*, 2012, **5**, 6990–7000.
41. A. R. C. Morais, A. M. da Costa Lopes and R. Bogel-Lukasik, *Chem. Rev.*, 2015, **115**, 3–27.
42. K. L. Toews, R. M. Shroll, C. M. Wai and N. G. Smart, *Anal. Chem.*, 1995, **67**, 4040–4043.
43. S. Takenouchi and G. C. Kennedy, *Am. J. Sci.*, 1964, **262**, 1055–1074.
44. R. P. Overend, E. Chornet and J. A. Gascoigne, *Philos. Trans. R. Soc., A*, 1987, **321**, 523–536.
45. H. L. Chum, D. K. Johnson, S. K. Black and R. P. Overend, *Appl. Biochem. Biotechnol.*, 1990, **24–5**, 1–14.
46. G. P. van Walsum, *Appl. Biochem. Biotechnol.*, 2001, **91–93**, 317–329.
47. G. P. Van Walsum, M. Garcia-Gil, S.-F. Chen and K. Chambliss, *Appl. Biochem. Biotechnol.*, 2007, 301–311.
48. R. C. McWilliams and G. P. van Walsum, *Appl. Biochem. Biotechnol.*, 2002, **98**, 109–121.

49. F. Carvalheiro, T. Silva-Fernandes, L. C. Duarte and F. M. Gírio, *Appl. Biochem. Biotechnol.*, 2009, **153**, 84–93.
50. S. P. Magalhães da Silva, A. R. C. Morais and R. Bogel-Lukasik, *Green Chem.*, 2014, **16**, 238–246.
51. A. R. C. Morais, A. C. Mata and R. Bogel-Lukasik, *Green Chem.*, 2014, **16**, 4312–4322.
52. T. Rogalinski, T. Ingram and G. Brunner, *J. Supercrit. Fluids*, 2008, **47**, 54–63.
53. J. W. King, K. Srinivas, O. Guevara, Y. W. Lu, D. F. Zhang and Y. J. Wang, *J. Supercrit. Fluids*, 2012, **66**, 221–231.
54. G. Brunner, *J. Supercrit. Fluids*, 2009, **47**, 373–381.
55. R. T. Dhamdere, K. Srinivas and J. W. King, Carbochemicals Production from Switchgrass using Carbonated Subcritical Water at High Temperatures, in 10th International Symposium on Supercritical Fluids, San Francisco, 2012.
56. M. Stamenic, I. Zizovic, R. Eggers, P. Jaeger, H. Heinrich, E. Roj, J. Ivanovic and D. Skala, *J. Supercrit. Fluids*, 2010, **52**, 125–133.
57. T. Gamse, R. Marr, F. Froschl and M. Siebenhofer, *Sep. Sci. Technol.*, 1997, **32**, 355–371.
58. K. B. Olanrewaju, Reaction kinetics of cellulose hydrolysis in subcritical and supercritical water, University of Iowa, 2012.
59. H. L. Clever, *J. Chem. Educ.*, 1968, **45**, 231.
60. D. A. Cantero, M. D. Bermejo and M. J. Cocero, *Bioresour. Technol.*, 2013, **135**, 697–703.
61. M. Sasaki, K. Goto, K. Tajima, T. Adschiri and K. Arai, *Green Chem.*, 2002, **4**, 285–287.
62. X. Yan, F. Jin, K. Tohji, T. Moriya and H. Enomoto, *J. Mater. Sci.*, 2007, **42**, 9995–9999.
63. W. L. Marshall and E. Franck, *J. Phys. Chem. Ref. Data*, 1981, **10**, 295–304.
64. M. Uematsu and E. Frank, *J. Phys. Chem. Ref. Data*, 1980, **9**, 1291–1306.
65. H. Pińkowska, P. Wolak and A. Złocińska, *Biomass Bioenerg.*, 2011, **35**, 3902–3912.
66. F. M. Relvas, A. R. C. Morais and R. Bogel-Lukasik, *J. Supercrit. Fluid.*, 2015, **99**, 95–102.
67. G. Brunner, *Gas Extraction. An Introduction to Fundamentals of Supercritical Fluids and the Applications to Separation Processes*, Springer, Berlin, 1994.

CHAPTER 7

Efficient Transformation of Biomass-derived Compounds into Different Valuable Products: A "Green" Approach

MAYA CHATTERJEE,*[a] TAKAYUKI ISHIZAKA[a] AND HAJIME KAWANAMI*[a,b]

[a] Microflow Chemistry Group, Research Institute for Chemical Process Technology, AIST Tohoku, 4-2-1 Nigatake, Miyagino-ku, Sendai 983-8551, Japan; [b] CREST Japan Science and Technology (JST), 4-1-8 Honcho, Kawaguchi 332-0012, Saitama, Japan
*Email: c-maya@aist.go.jp; h-kawanami@aist.go.jp

7.1 Introduction

The term "biomass" refers to biological materials derived from living organisms, typically the cellular structure of a plant or animal.[1] It is the most abundant, inexpensive and renewable energy resource, which reduces environmental hazard. Lignocellulosic biomass is a generic term for dry plant-based material, mainly consisting of three components: cellulose, hemicellulose and lignin. Depending on the source, the component composition varies. In general, lignocellulosic biomass contains 35–50% cellulose, 20–35% hemicellulose and 10–25% of lignin. Cellulose is the major components of lignocellulosic biomass, consisting of mainly glucose units; glucopyranosil

Green Chemistry Series No. 48
High Pressure Technologies in Biomass Conversion
Edited by Rafał M. Łukasik

Published by the Royal Society of Chemistry, www.rsc.org

monomers linked by 1,4-β-glycosidic bond and form a nice linear polymer, unlike starch. Due to the linear conformation, the numerous cellulose strands can packed into crystalline cellulose fibrils.[2] Those cellulose fibrils are stabilized through the van der Waals interaction.[3] Cellulose is a tough component of lignocellulosic biomass because of the high degree of polymerization and the polymer chains having low flexibility due to their intermolecular hydrogen bonding.[4] The next module is hemicellulose, which is a group of polysaccharides with a degree of polymerization of only 200. Common sugars found in hemicellulose are hexose, pentose, glucose, mannose, galactose as well as xylose and arabinose. Hemicellulose is a mixed polymer and the main structural feature is a chain of $(1\rightarrow4)$-linked β-D-xylopyranose units.[5] Lignin is another component present in the cell wall of the plant, which acts as glue between fibres in wood microstructure and protects it internally.[6–8] After cellulose it is the most abundant, water insoluble three-dimensional network polymer composed of phenylpropane units and the subunits are *p*-coumaryl alcohol, coniferyl alcohol and sinapyl alcohol, those are essentially varied by the number of substituted methoxy groups on the aromatic rings.[9,10] Those units preferred to couple through C–C bond and ether linkages essentially results β-O-4, β-1, α-O-4, 4-O-5 *etc.* The exact structure of lignin is difficult to predict and accordingly the exact molecular weight of natural lignin is unsure. The presence of many polar groups resulting strong intermolecular and intramolecular hydrogen bonding making the structure tough with the combustion energy of 26.6 kJ g^{-1}, the highest among all polymeric compounds containing C, H and O.

Considering the interesting structural aspect, lignocellulosic biomass is one of the few resources showing potential to meet the challenges to produce various chemicals in more energy efficient and environmentally sustainable way. In this regard, the United States and the European Union established a "biorefinery" concept in 1980. The main goal of a "biorefinery" is to provide higher-value products including fuel, power and chemicals with no waste, by integrating biomass components. The overall challenges with biomass derived feedstock is to convert them into simple and well-defined molecules called "platform molecules".[11,12] In 2004, the DOE (Department of Energy) short-listed 12 platform chemicals. Among them, furfural, 5-hydroxymethylfurfural (5-HMF), levulinic acid and γ-valerolactone are of special interest.[13]

5-HMF is considered as one of the most versatile platform molecules synthesized by acid catalyzed dehydration of sugar, mainly hexose. A comprehensive review of the synthesis chemistry of 5-HMF and its derivative was provided by Lewkowski.[14] In the past few years, an enormous increase in the number of publications, including several reviews considering different features of 5-HMF has been observed.[15–20] From all these literatures, the importance of 5-HMF as a bio-based platform molecule is clearly evident. The importance of 5-HMF lies with its interesting multi-functionality structure, which contains two different functional groups along with a furan ring, which can undergo different types of reactions such as hydrogenation of the C=O bond, furan ring, hydrogenolysis of C–O bond, rearrangement, dissociation, polymerization *etc.* and can be converted into different fuel and non-fuel related compounds.

The hydrolysis and dehydration of xylan contained in hemicellulose part of the lignocellulosic biomass produce another promising platform compound called furfural, which can also generate a potential biofuel 2-methylfuran (2-MF), through the hydrogenation and hydrogenolysis over metal supported catalyst. Upgradation of furfural to biofuel require deoxygenation and the most studied routes are hydrogenation, rearrangement and C–C coupling.[21]

Selective hydrogenolysis of tetrahydrofurfuryl alcohol (THFA) to 1,5-pentanediol (1,5-PD), represents an important class of reaction, because diols are widely used as a monomer for the production of polyester and polyurethane. An excellent review covering both homogeneous and heterogeneous catalysis focusing on the hydrogenolysis of C–O bond to the formation of diol in different organic solvent was provided by Schlaf.[22]

To develop an efficient transformation process of biomass-derived multi-functionality furan compounds into value-added chemicals has technological importance for a sustainable future. Considering this, right choice of the solvent is extremely important in chemical synthesis due to its significant effects on the energy consumption as well as on the E-factor (mass of waste/mass of desired product) of the related process.[23]

The quest for a green reaction medium as a replacement of volatile organic solvent is one of the dynamic parts of fundamental research in chemistry, chemical engineering and their related areas as there are multiple factors to consider.[24] Undoubtedly, $scCO_2$ and H_2O are the most widely used green solvents in organic synthesis. Clark and Tavener used a scoring system to define the general level of greenness and rated CO_2 and H_2O as 18/25 and 19/25, respectively.[25]

Supercritical CO_2 ($scCO_2$) is an environmentally friendly solvent because it does not increase VOC's level in the atmosphere, is non-toxic, non-flammable and easily available. The phase diagram of CO_2 is characterized by the critical temperature = 31.0 °C, critical pressure = 7.38 MPa and the corresponding critical density of 0.466 $g\,mL^{-1}$ (Figure 7.1). The most intriguing and gripping criteria of $scCO_2$ to qualify for the greatest-studied green reaction medium is process benefit, connected with the rare occurrence of by-products owing to side reactions, absence of solvent residue and facile product separation (cost-effective; related to the separation and purification steps). In the case of chemical reactions involving gases, $scCO_2$ is the most potential medium due to the complete miscibility of reactant gasses like hydrogen, oxygen *etc.* Simple tuning of the reaction parameters such as pressure and temperature can accelerate the reaction rate and dictate the product selectivity. Over the past few years, heterogeneous catalysts were used for different types of reactions in $scCO_2$, often with higher reaction rates and different product distributions, as well as high selectivity, in comparison with those in conventional organic solvents.[26–31] Although, $scCO_2$ is considered as a relatively inert solvent, chemical interaction of

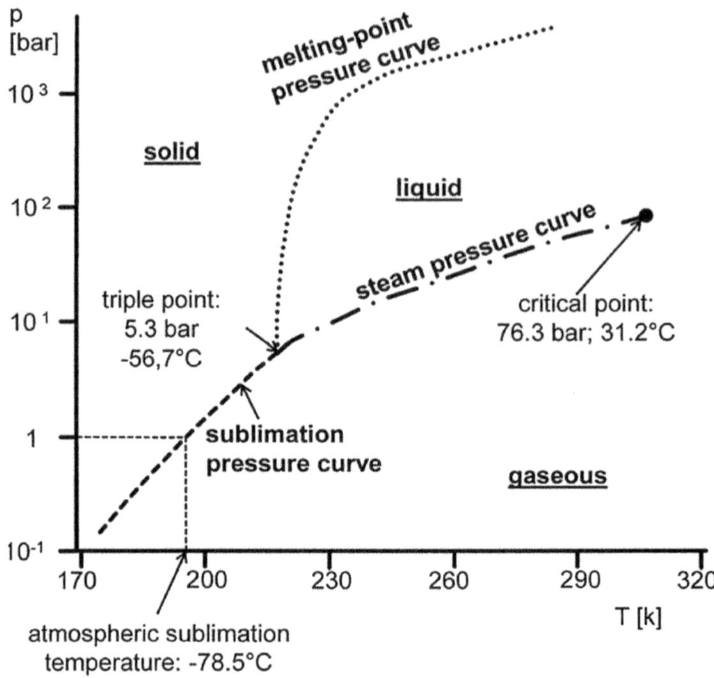

Figure 7.1 Phase diagram of CO_2.
Reproduced from ref. 84 with permission from The Royal Society of Chemistry.

CO_2 with specific substrates sometimes offer great potential to improve selectivity.[32,33] One of the greatest advantages of using $scCO_2$ as reaction medium is to overcome the mass transfer limitation frequently occur between the gaseous reactant and the liquid solvent medium.

While talking about the alternative solvent, H_2O is the ultimate choice, since its widely availability, non-toxic, non-flammable, inert and cheapest solvent in this world. H_2O has ability to establish large three dimensional network of hydrogen bond because of its polarity, which can influence the reactivity of substrates.[34] Furthermore, the highest heat capacity of H_2O makes it enable of conducting exothermic reactions. H_2O can be utilized for the reaction where additives are needed because of its interesting phenomenon of "salting-in" and "salting-out" effects. Considering the biphasic system, organic compounds can be easily separated by simple phase separation.[35,36]

From the above discussion $scCO_2$ and H_2O have lots of benefits to act as reaction medium for organic synthesis in an environmentally safe way. However, besides all these attractive properties of $scCO_2$ and H_2O, both of them have some limitations. For example, $scCO_2$ is a relatively poor solvent. Although it is an inert solvent, sometimes it interacts strongly with powerful nucleophilic reactants due to the high quadrupole moment. Therefore, it cannot be used for some reactions involving strong nucleophiles, like amines.

H_2O also has certain restrictions as a solvent for the reactions with organic substrates. It encounters the problem of H_2O sensitivity of substrates, catalysts as well as the generated intermediates. In addition, an H_2O-soluble product requires distillation for product separation, and consequently enhances the energy consumption of the process; it is also difficult to treat the contaminated waste H_2O.

Therefore, to avoid those drawbacks a combination of $scCO_2$–H_2O can be used together, which offer great potential as reaction medium for different chemical transformations. The solvation power of CO_2 and H_2O are complimentary and forms an ideal reasonable solvent environment for solubilising a wide range of molecules.[37] The accelerative effect of H_2O can be readily assessed when use along with $scCO_2$. Furthermore, product separation and recyclability of catalyst might be improved.

Heterogeneous catalysts, especially metal nanoparticles supported on various materials are widely accepted as a successful catalyst for different types of organic synthesis. Here, mesoporous silica (MCM-41) activated carbon (C) and alumina (Al_2O_3) are the three support materials for metal nanoparticles considered in this work. In 1992, Beck *et al.* discovered that the mesoporous material (MCM-41) seems to be an ideal host for metal nanoparticles because of the very high surface area and tuneable pore size.[38] Typically, mesoporous materials are synthesized by a "liquid-crystal" templating method, where micelles of a cationic surfactant (alkyltrimethylammonium) acted as a template and directed the formation of mesoporous structure *via* electrostatic interaction with anionic silica precursor. Generally, MCM-41 possesses ordered hexagonal structure with large surface area of 1500 $m^2 g^{-1}$, high thermal stability and the pore diameter can be tailored between 2–10 nm.[39,40] Although MCM-41 is an inert support, great diversity can be generated simply by the substitution of Si with different heteroatoms like Al, B, or Ga to incorporate acid sites in the support.

The processing of 5-HMF, furfural and THFA in $scCO_2$ or $scCO_2$–H_2O medium, employing heterogeneous metal catalysts under a mild reaction condition and the detail description of the transformation through different types of reaction such as hydrogenation, hydrogenation/hydrogenolysis *etc.* is discussed in this chapter.

7.2 Experimental Methods

7.2.1 General Method for *in situ* Synthesis of Metal Nanoparticles Supported MCM-41

Synthesis of metal nanoparticles supported MCM-41 was carried out by an *in situ* method using a molar gel composition of 1 TEOS: 0.45 Na_2O: 0.12 CTAB: 118 H_2O. Typically, cetylytrimethylammonium bromide (CTAB), which acts as a template and sodium hydroxide (NaOH) was added to deionized water. The solution was stirred continuously at room temperature until the template gets dissolved. After that, tetraethylorthosilicate (TEOS), as a silica source, is introduced slowly under stirring and continued for another hour. In the case of substitution of Si by trivalent metal ions, required amount of the corresponding metal salt along with TEOS were added to the gel at this stage. For *in situ* generation of metal nanoparticles, 1 wt% solution of the desired metal salt solution has been introduced into the gel after the addition of TEOS under the continuous stirring for another 2 h. The gel was then autoclaved at 140 °C for 48 h. The solid product was filtered from the supernatant, washed thoroughly with deionized water, followed by oven drying at 60 °C. To remove the template, the as-synthesized material was calcined at 550 °C for 8 h in air and then tested as a catalyst for different reactions.

7.2.2 Catalyst Characterization

7.2.2.1 X-Ray Diffraction (XRD)

A typical XRD pattern of Si-MCM-41 consisted of a characteristic low angle reflection (100) with smaller Bragg peaks[38] and is shown in Figure 7.2(a). The introduction of metal particles results in the characteristic peaks of different metal ions in the higher angle region ($2\theta = 30°$ to $70°$) such as for calcined Pd/MCM-41, five diffractive peaks at $2\theta = 33.8$, 42.0, 54.8, 60.7 and 71.4 corresponding to PdO was detected, but after the reaction PdO was converted to metallic Pd (Figure 7.2(b)). The XRD pattern of the calcined Rh/MCM-41 exhibits a peak at $2\theta = 34.3°$ indicating the presence of Rh_2O_3 particles with orthorhombic symmetry[41] (Figure 7.2(c)).

7.2.2.2 Transmission Electron Microscopy (TEM)

Figure 7.3(a) shows a TEM image of the calcined Si-MCM-41 used as a support material for different nanoparticles. Highly ordered hexagonal structure of Si-MCM-41 can be viewed from the Fast Fourier Transformation (FFT) pattern (inset) after the removal of template by calcination. In most of the cases, MCM-41 maintained its structural order even after the incorporation of metal nanoparticles as evident from Figure 7.3(b) (FFT pattern as an inset). Moreover, presence of metal nanoparticles can be seen as black dots, distributed throughout the matrix.

7.2.3 Catalytic Activity

All the reactions were conducted in a 50 mL stainless-steel batch reactor and the reaction setup is shown in Figure 7.4.

An appropriate amount of catalyst and the reactant were introduced into the reactor and placed in an oven with fan heater to maintain the desired temperature. The reactor was purged with CO_2 to remove air. After the temperature had stabilized, hydrogen was introduced into the reactor followed by liquid CO_2 using a high-pressure liquid pump (JASCO) and then

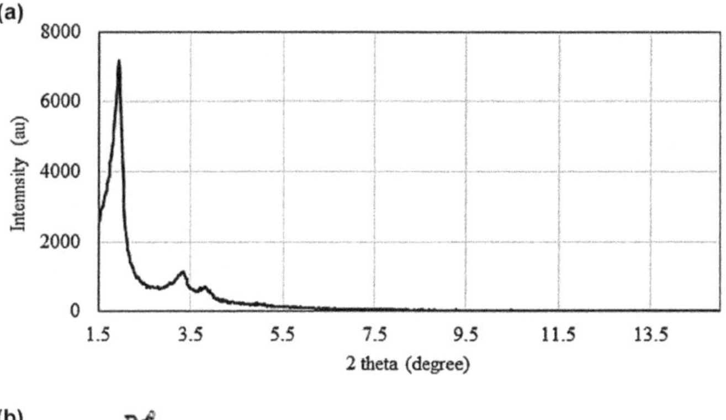

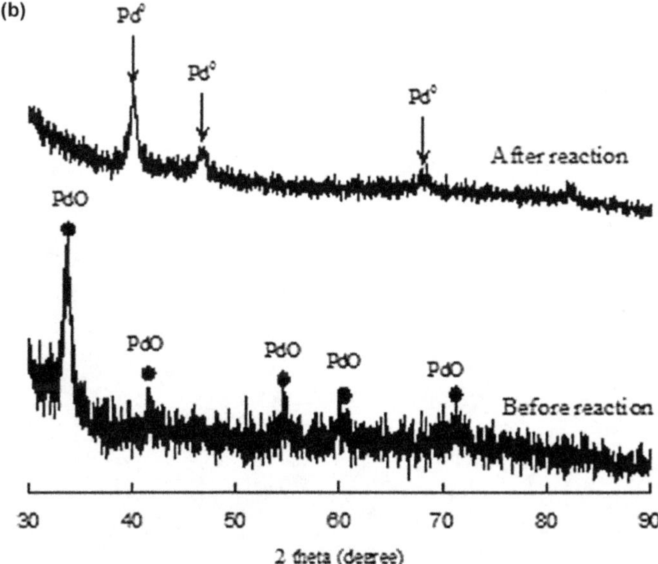

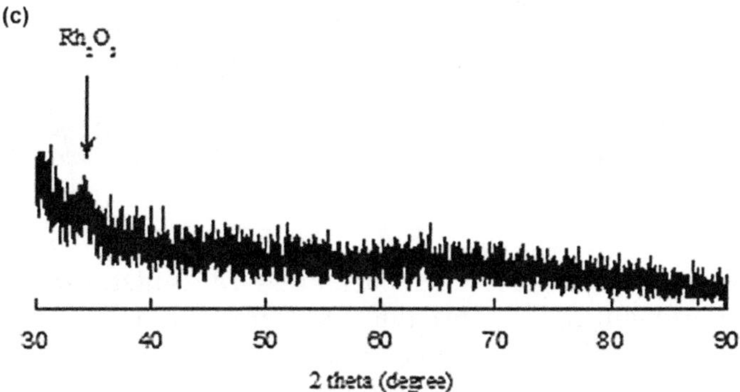

Figure 7.2 Low-angle and high-angle XRD pattern of (a) Si-MCM-41 and (b) Pd in Pd/MCM-41, respectively. (c) Rh_2O_3 nanoparticles.
Reproduced from ref. 85 and 86 with permission from The Royal Society of Chemistry.

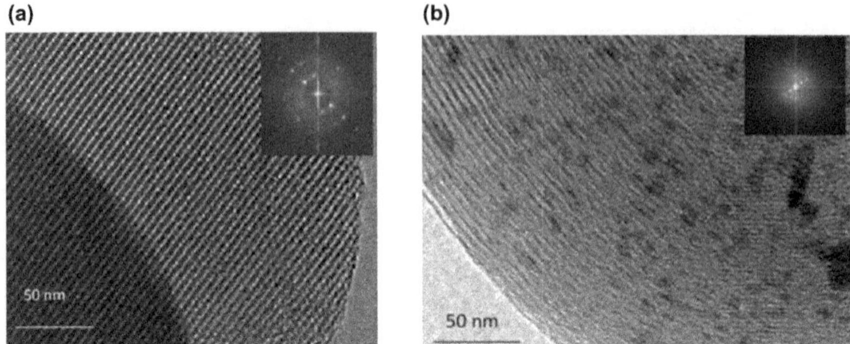

Figure 7.3 Representative TEM images of (a) Si-MCM-41 and (b) metal nanoparticles supported on MCM-41.
Reproduced from ref. 86 with permission from The Royal Society of Chemistry.

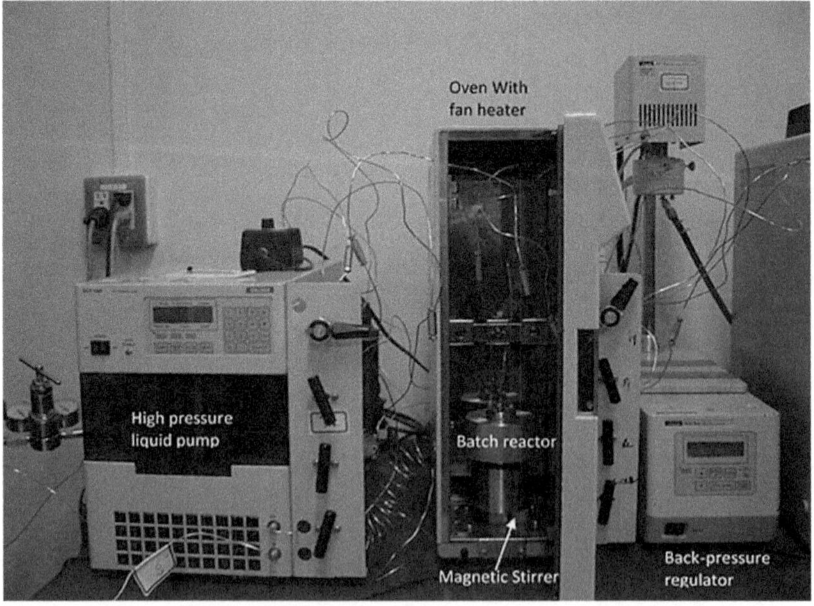

Figure 7.4 Batch reactor setups.

compressed to required pressure. A back-pressure regulator maintained the system pressure during the reaction. The reaction mixture was stirred continuously with Teflon-coated magnetic bar throughout the reaction. After the reaction, the reactor was placed in an ice-bath to cool, and then depressurized carefully. The liquid product was separated from the catalyst by filtration and identified by NMR or by GC-MS (Varian Saturn 2200). Quantitative analysis was conducted using a GC (HP 6890) equipped with capillary column and flame ionization detector. For all results reported, the selectivity was calculated by the following expression: $S_i = C_i/\sum Cp$, where C_i is the concentration of the product 'i' and $\sum Cp$ is the total concentration of the product.

7.2.4 Phase Behaviour Studies

Before conducting the reaction, it is necessary to check the solubility of the substrate in $scCO_2$ and the corresponding phase behaviour under the applied reaction condition. For this purpose, a 10 mL high-pressure view cell fitted with a sapphire window was used separately. The view-cell was placed over a magnetic stirrer for stirring the content and connected to a back-pressure controller, to regulate the pressure inside the cell. In addition, a temperature controller was also incorporated to maintain the required temperature. A calculated amount of substrate was introduced into the cell followed by the temperature stabilization of the system. The reactant gas (H_2) and CO_2 were introduced into the cell in the same manner as that of in the batch reactor as mentioned previously. Phase behaviour of CO_2, H_2 and the substrate was monitored through a video camera and the images of each stage were recorded.

7.3 Results and Discussion

7.3.1 Catalytic Strategies to Process 5-HMF into Fuel

In recent years, research on alternative fuel from biomass has received considerable attention, since the increasing concerns about fossil fuel depletion, global warming and environmental pollution. Growing demand of energy, especially in the transportation sector expects to increase significantly in near future due to the strong economic growth of some developing countries. Biomass is one of the suitable renewable energy resources, which offers a great potential through the long-term storage of carbon and also considered a part of a closed carbon cycle. It can reduce the greenhouse gas emission in comparison with fossil fuel, which produces 23 billion tonnes of CO_2 per year. The utilization of biomass has an extreme importance for sustainable developments because it contains high carbon to oxygen ratio.[42–45] The production of hydrocarbon from biomass has several advantages such as compatibility with existing system because of the similarity with petroleum counterpart, high heating value, fewer emission, immiscibility with water and more flexibility.

7.3.1.1 5-HMF to Alkane

Several methodologies were developed for the generation of liquid fuel from biomass. Huber *et al.* demonstrated the possible formation of light alkane from biomass-based sorbitol by aqueous phase reforming (APR) process, which is advantageous because of the easy separation of alkanes from water and also energy efficient.[46] The obtained product mixture was mainly contained hexane along with other lighter alkanes ranging from pentane to methane and the total yield of alkane was reached to 60%. Although, overall conversion of sorbitol to alkane was satisfactory the largest alkane obtained was hexane; high volatility of this compound makes it a poor candidate as a fuel additive. The strategy was then improved to convert biomass into liquid alkanes of carbon number ranging from C7 to C13. The modified process was consisted of aldol condensation between HMF and acetone in the presence of basic Mg–Al oxide catalyst to increase the carbon chain followed by the hydrogenation of the condensation product using Pd/Al_2O_3 at 120 °C and 6.5 MPa of hydrogen pressure in a batch reactor. The final step of alkane formation was dehydration/hydrogenation conducted at 250–265 °C, 5–6.2 MPa of hydrogen pressure in the presence of $Pt/SiO_2–Al_2O_3$ catalyst. Depending on the condensation product (monomer or dimer) a mixture of alkane was obtained.[11] However, one of the main problems associated with the deoxygenation of biomass-derived oxygenates is the requirement of large amounts of hydrogen in the hydrogenation step.

Therefore, an attempt was made to develop a simple method for generation of a single alkane from 5-HMF in $scCO_2$ medium and to find a possible solution of excess hydrogen usage during hydrogenation, which has a strong impact on the operational cost (Scheme 7.1). The developed process was also

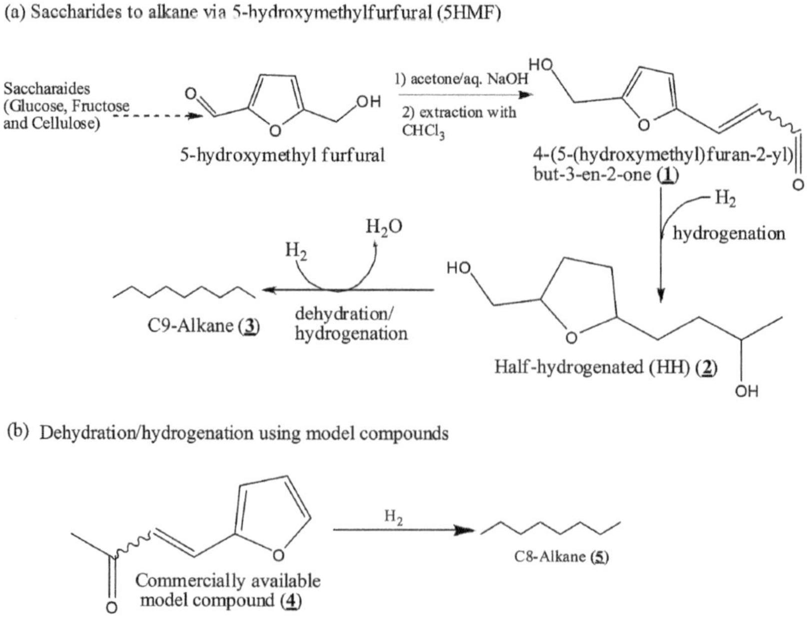

(a) Saccharides to alkane via 5-hydroxymethylfurfural (5HMF)

Saccharaides (Glucose, Fructose and Cellulose)
5-hydroxymethyl furfural
1) acetone/aq. NaOH
2) extraction with $CHCl_3$
4-(5-(hydroxymethyl)furan-2-yl) but-3-en-2-one (1)
H_2 hydrogenation
Half-hydrogenated (HH) (2)
H_2 H_2O
dehydration/ hydrogenation
C9-Alkane (3)

(b) Dehydration/hydrogenation using model compounds

Commercially available model compound (4)
H_2
C8-Alkane (5)

Scheme 7.1 A possible reaction pathway of conversion of HMF to linear alkane under the studied reaction condition.
Reproduced from ref. 85 with permission from The Royal Society of Chemistry.

consisted of (i) aldol condensation between 5-HMF and acetone and (ii) hydrogenation of aldol condensation product in $scCO_2$. Scheme 7.1(a) depicts a possible reaction path of the transformation of 5-HMF into a single alkane. In the first step, base catalyzed aldol condensation between 5-HMF and acetone was occurred in 25–40 °C in the presence of NaOH without any catalysts. The light yellow intermediate was extracted with chloroform followed by the purification by chromatographic column separation and evaporation of the solvent. The resulted intermediate was then subjected to hydrogenation in the presence of a metal catalyst at 80 °C, 14 MPa and 4 MPa of CO_2 and H_2 pressure, respectively. For the sake of simplicity, a model compound **4**, which is very close to the intermediate compound obtained from the aldol-condensation between HMF and acetone, was used to optimize the reaction condition (Scheme 7.1(b)). Table 7.1 presents a catalyst screening summary of the hydrogenation of model compound **4**. Results show that Rh and Pd gave complete conversion of **4** (Entries 1–5 and 7), but Pt was totally inactive (Entry 6). Interestingly, alkane formation was restricted only on the Pd catalysts supported on Al containing materials (Entries 1–3) and very high yield of alkane (>99%) was detected only on the Pd/Al-MCM-41 (Entry 1). Hence, Pd/Al-MCM-41 used for the further optimization of different reaction parameters. Regarding the CO_2 pressure, it has a tremendous effect on the conversion and also on the product selectivity. Conversion of the compound **4** was enhanced from 70 to >99% with increasing the CO_2 pressure from 7 to 10 MPa (Figure 7.5). At 7 MPa only 7% of alkane was formed and the major product detected with 95% selectivity was the furan ring hydrogenated product (termed as half-hydrogenated product (HH); compound **5**). However, the selectivity of targeted alkane was started to increase with the pressure and reached a maximum of >99% above 10 MPa. This result suggested that at the lower pressure of CO_2, the amount of H_2 present in the reaction system was

Table 7.1 Catalyst screening for hydrogenation and dehydration/hydrogenation of model compound **4**. Adapted from ref. 85 with permission from The Royal Society of Chemistry.[a]

Entry	Catalyst	Conv. (%)	Selectivity (%)	
			HH (**5**)	Alkane (**6**)
1	Pd/Al-MCM-41	>99.0	0.0	>99.0
2	Pd/Al_2O_3	>99.0	37.5	62.5
3	$Pd/SiO_2–Al_2O_3$	>99.0	9.2	90.8
4	Pd/C	>99.0	100.0	0.0
5	Pd/zeolite Y	68.2	0.0	>99.0
6	Pt/C	0.0	0.0	0.0
7	Rh/C	>99.0	72.5	0.0
8[b]	Pd/Al-MCM-41	>99.0	0.0	>99.0

[a]Reaction conditions: temperature = 80 °C, p_{CO_2} = 14 MPa, p_{H_2} = 4 MPa, reaction time = 20 h. Entries 2, 4, 6, and 7 are commercial catalysts from Aldrich; Entries 3 and 5 are ion exchanged Pd on commercial support; Entry 7 = 20% unknown product; HH = furan ring hydrogenated product.

[b]Substrate **1** and product is alkane **3**.

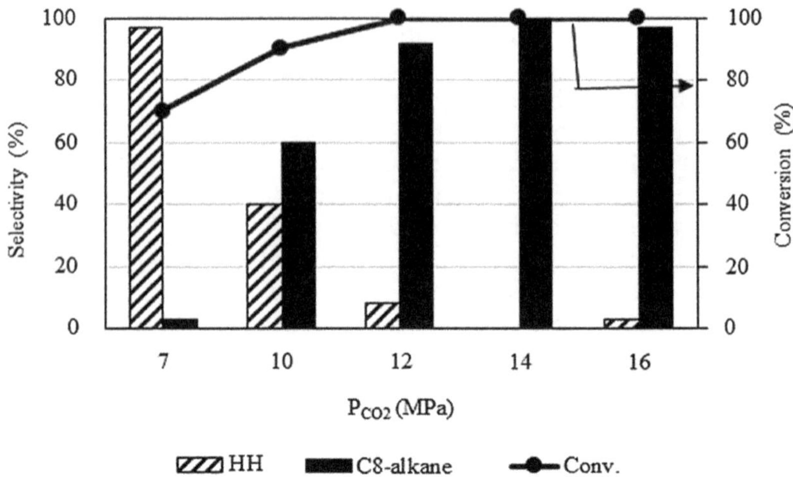

Figure 7.5 Effect of CO_2 pressure on the transformation of model compound **4** from Scheme 7.1. Reaction conditions: temperature = 80 °C, reaction time = 20 h, p_{H_2} = 4 MPa; HH = compound **5** and C8-alkane = compound **6** (Scheme 7.1(b)).
Reproduced from ref. 85 with permission from The Royal Society of Chemistry.

sufficient enough for the generation of HH, but could not pursue the ring opening reaction.

From the above results, it is confirmed that the model compound **4** was turned into HH in the hydrogen deficient condition.

A time course of the reaction was monitored at the selective pressure of 14 MPa, which is required to achieve high selectivity of alkane. The conversion of compound **4** as a function of reaction time indicate that in the beginning, furan ring was hydrogenated to HH with 99% selectivity within the reaction time of 1 h. However, as the time progresses the selectivity of alkane (**6**) started to increase and subsequently the selectivity of HH was dropped. At least the reaction time of 15–20 h was required to achieve highest conversion of **4** and the selectivity of desired alkane under the studied reaction condition. From the present observation, HH could be considered as an intermediate in the conversion of compound **4** to alkane. In the presence of a metal catalyst, the hydrogenation of C=C and C=O of the HH occurred, followed by the dehydration/hydrogenation to result in the liquid alkane.[47] Notably, the hydrogenation of the furan ring of compound **4** was faster than other reaction but the transformation of the HH to alkane required longer reaction time because it involves the hydrogenolysis of C–O, which follows two steps: dehydration and hydrogenation, facilitated by the acid and metal site, respectively.[48] The positive effect of temperature on the conversion of 5-HMF and the selectivity also explains that the large amount of energy is needed for the opening of the furan ring as there was exceptionally low activity at the temperature of below 80 °C.

The same deoxygenation method was also attempted on sulfur and nitrogen containing compounds to remove those heteroatoms. These two elements are well-known as the most notorious and undesirable contaminants in crude petroleum. Nitrogen compounds mainly pyridine, quinolone, amine and their alkyl substituted derivative and the sulphur compounds such as thiols, sulfide, thiophene *etc.* are present in crude oil. The oil refining process is hampered by nitrogen compounds, which causes catalyst poisoning as well as colour formation in the final product.[49] During combustion, sulfur is converted to sulfur dioxide and contributes to environmental pollution. Hence, imidazole and thiophene were chosen as the two compounds for the removal of nitrogen and sulfur, respectively. Under the same reaction conditions, 94% of imidazole was converted, but the resultant product/products remain unidentified because of their volatile nature. Unfortunately, when thiophene was used as a substrate, the catalyst was completely inactive, under the present reaction conditions.

In summary, the developed method exploited $scCO_2$ as an effective reaction medium for the successful conversion of 5-HMF to liquid alkane in a mild reaction condition. As hydrogen is miscible in $scCO_2$, relatively lower pressure of hydrogen worked perfectly to the formation of alkane. However, it remains inactive for other heteroatoms containing compounds, which required large amounts of energy and continuous effort to extend this method.

7.3.2 5-HMF to 2,5-Dimethylfuran (DMF)

Biofuels are a part of a growing global industry to reduce the need for fossil fuels. Currently, biofuels are mainly produced from the food crops such as corn, wheat, sugar cane *etc.* by fermentation followed by distillations and a topic of global debate. By far the most dominating biofuel on the market is ethanol. In 2014, 14.3 billion gallons of ethanol was produced in US, which is 57% of global production. Thus, in the US, ethanol is used as an additive and 90% of gasoline mixed with 10% of ethanol marked as E10 help to reduce CO and smog causing emission.[50] However, ethanol as biofuel suffers from a number of drawbacks such as low energy density of 23 $MJ\,L^{-1}$, high volatility (boiling point 78.4 °C) and complete miscibility with water (problematic for cold weather). In addition, other concerns are cost of production, requirement of large arable land and the corrosive nature of ethanol (typical storage or engine tank need modification). As an alternative, a second-generation biofuel DMF was introduced, obtained by the deoxygenation of 5-HMF. Hydrogenation of the carbonyl group of 5-HMF results in dihydroxymethyl furan as an intermediate and the subsequent hydrogenolysis of the C–O bond of the resultant compound produces DMF. Compared to ethanol, DMF has high boiling point (92–94 °C), insoluble in water, stable in storage, very high research octane number (RON) of 119° and most attracting feature is DMF possesses a high volumetric energy density of 31.5 $MJ\,L^{-1}$, which is 40% higher than ethanol.[51] Compared to fermentation process of ethanol generation, DMF requires one-third of the energy in its evaporation step of production. All these criteria make DMF as a most appropriate candidate of an alternative fuel. Recently, DMF has shown satisfactory results while tested on a single-cylinder gasoline engine in terms of emission and combustion.[52]

The breakthrough discovery of DMF from biomass by Leshkov *et al.* paves the way to develop another option in the biofuel industry. They presented a catalytic strategy for the production of DMF from a biomass based carbohydrate called fructose. This cellulose based compound was converted to DMF using a bimetallic Cu–Ru/C catalyst in 1-butanol at 220 °C and 0.68 MPa of hydrogen. The yield of DMF was 71% within a reaction time of 10 h.[51] The selected catalyst deactivated slowly on stream, but could be regenerated through the treatment of flowing hydrogen. Binder and Raines also reported 9% yield of DMF from corn stover in the presence of a $CrCl_3$–HCl catalyst.[53] In a separate approach, Chidambaram and Bell obtained a maximum yield of 13% with the highest HMF conversion of 19% within 1 h in the ionic liquid medium (1-ethyl-3-methylimidazolium chloride ([emim][Cl])) at 6.2 MPa of hydrogen and catalysts were Pd, Rh and Ru supported on activated carbon. The lower activity in the ionic liquid medium was explained due to the low solubility of hydrogen in the medium as well as the shorter reaction time.[54] The requirement of the high hydrogen pressure due to the low solubility of hydrogen makes the process energy intensive. Furthermore, catalytic transfer hydrogenation (CTH) was also employed in which instead of molecular hydrogen, different reaction medium played dual role of solvent and as a hydrogen source. Although the yield of DMF was improved to 60–80% depending on the medium used, associated disadvantages are the formation of by-products or co-products and harsh reaction conditions.[55–63] As far as the yield of DMF was concerned, Thananatthanachon and Rauchfuss achieved a maximum yield of 95% while refluxing HMF in the presence of H_2SO_4, and formic acid for 15 h at 120 °C on Pd/C catalyst,[64] but acidic co-catalyst was difficult to separate, which generates undesirable by-products. Several other strategies were also developed to achieve high yield of DMF directly from 5-HMF or carbohydrate.[65–70] At a glance, most of the described methods involve the utilization of a complex catalytic system and harsh reaction conditions (high temperature and hydrogen pressure) to achieve the best yield of DMF. Hence, main challenge is the development of an appropriate method in a simple and environmentally friendly way to avoid the complexity of the process.

To design a method for the hydrodeoxygenation of HMF to DMF, the conversion of HMF was attempted in $scCO_2$ medium using a commercial 5% Pd/C catalyst (Scheme 7.2).[31] To search for a potential catalyst, a series of metals (Pt, Pd, Rh and Ru) supported on MCM-41, C and Al_2O_3 were tested for the transformation of HMF to DMF under the optimized reaction condition (temperature = 80 °C, reaction time = 2 h, H_2O = 1 mL, $p_{CO_2} = 10$ MPa and $p_{H_2} = 1$ MPa). Based on the support materials, the yield of DMF was different; C supported catalysts showed best activity in terms of conversion and

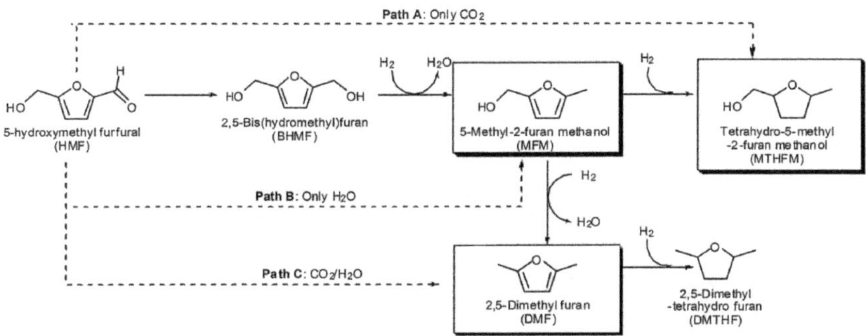

Scheme 7.2 Solvent dependent reaction scheme of the transformation of 5-HMF to DMF under the applied reaction condition.
Reproduced from ref. 87 with permission from The Royal Society of Chemistry.

the selectivity of DMF. The conversion was significantly dropped for Al_2O_3 and MCM-41 supported catalysts and failed to produce DMF. It must be mentioned that Pd/C was highly active and forms DMF of very high selectivity (>99%) under the stated conditions.

Continuing with the Pd/C catalyst, an interesting effect of CO_2 pressure on the product distribution was detected. Three different products in the three different regions of CO_2 pressure (<10 MPa, 10 MPa and >10 MPa) were spotted. In the low-pressure region of <10 MPa, 57.8% tetrahydro-5-methyl-2-furanmethanol (MHTFM) was the major product, whereas at 10 MPa, DMF was successfully produced with >99% selectivity. When the pressure was increased further (>10 MPa), furan ring was saturated to form 2,5-dimethyltetrahydrofuran (DMTHF) (selectivity = 70%). To explain this interesting observation, the phase behaviour between the 5-HMF, H_2O (liquid phase) and CO_2–H_2 (gaseous phase), was conducted separately. Figure 7.6(a) to 7.6(d) depicts the images taken through a high-pressure view cell at 0, 4, 10 and 16 MPa CO_2 pressure. Depending on the applied CO_2 pressure, the images are significantly different. Figure 7.6(a) represents the initial condition (0 MPa) of the system containing 5-HMF, H_2O and solid catalyst. When 4 MPa of CO_2 was introduced, two phases containing H_2O–substrate (liquid) and CO_2–H_2 (gaseous) appeared as bottom and upper phase, respectively (Figure 7.6(b)). At 10 MPa, three distinctly visible phases are observed (i) substrate–H_2O/catalyst (bottom phase), (ii) liquid part (middle phase) and the (iii) CO_2–H_2 in a gaseous state (upper phase) (Figure 7.6(c)). The bottom and upper phases are expected to be similar with 4 MPa but the appearance of a third phase present in the middle make the scenario significantly different. However, at the high pressure of 16 MPa, a large portion of the view cell is occupied by liquid CO_2–H_2 as framed (Figure 7.6(d)). Hence, from the perspective of phase behaviour it appears that these three different situations albeit three different products formed at three different CO_2 pressure region. Interestingly, at lower pressures, the hydrogenolysis of only one hydroxymethyl group occurred and most

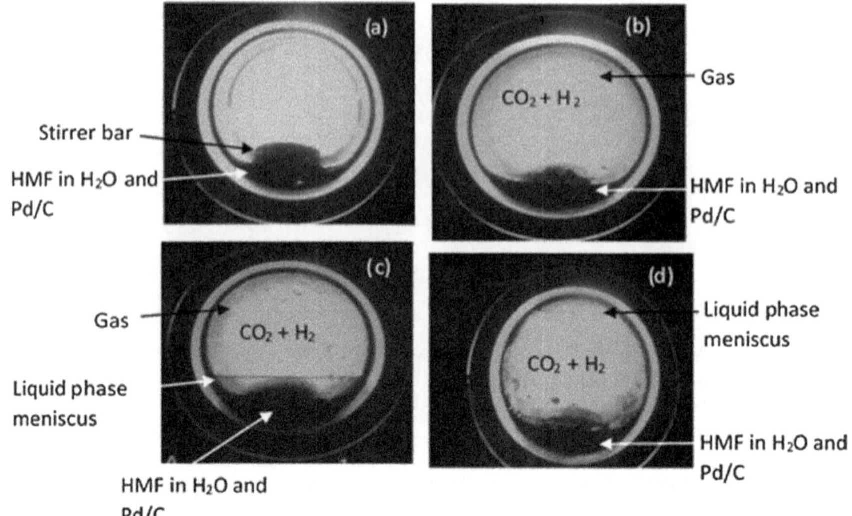

Figure 7.6 Phase observation through view cell during reaction (a) 0 (b) 4, (c) 10 and (d) 16 MPa of CO_2 pressure. Reaction conditions: temperature = 80 °C, reaction time = 2 h, $p_{H_2} = 1$ MPa.
Reproduced from ref. 87 with permission from The Royal Society of Chemistry.

strikingly the hydrogenation of conjugated C=C of the furan ring was taken place. It is well-evidenced that the hydrogenation of the conjugated C–C of furan ring is difficult, however, in the presence of Pd catalyst, hydrogenation of C=C bond of furan ring was preferred as documented during the hydrogenation of furan.[71] The transformation of 5-HMF to DMF was best described under the three phase condition occurred at 10 MPa in which the hydrogenolysis of two hydroxymethyl group was more facile rather than the furan ring hydrogenation. DMF is immiscible in H_2O, but dissolves in $scCO_2$, hence, once DMF was formed it might be readily get extracted and the highest selectivity was achieved. Furthermore, the DMTHF was observed in the system where CO_2 homogenize the mixture of DMF and hydrogen, leading to high concentrations of each reactant and consequently high DMTHF selectivity at the higher pressure of CO_2 (16 MPa).

It is well-known that hydrogen has a complete solubility in $scCO_2$, hence, the effect of hydrogen pressure on the catalytic activity was prominent. The conversion of HMF was changed from 47 to 100% as the pressure changes from 0.1 to 2 MPa and DMF was detected as a sole product. However, DMF was further hydrogenated to form DMTHF when the pressure reached above 2 MPa.

To identify the intermediate under the applied condition (temperature = 80 °C, $p_{CO_2} = 10$ MPa and $p_{H_2} = 1$ MPa), the reaction time was shortened to 5 min. Apart from DMF, no other product was traced and direct conversion of 5-HMF to DMF could be predicted. However, at a very low pressure of hydrogen (0.1 MPa), when the reaction was conducted for 5 min the conversion of 5-HMF was substantially dropped and the product mixture contain 2,5-bis(hydroxylmethyl)furan (BHMF) as a major compound, which inferred that the presence of the large amount of hydrogen causes a very fast reaction and the identification of the intermediate was difficult.

Another primary factor needs to be optimized is the reaction temperature and the effect is crucial when the reaction medium is $scCO_2$. The change in temperature not only influenced the conversion, but also significantly improved the selectivity of DMF (Figure 7.7). To compare the performance, the reaction temperature varied from 35 to 130 °C. The conversion of 5-HMF jumped from 30 to 100% with the change in temperature from 35 to 80 °C due to the kinetic effect. Again, at the low temperature of 35 °C, the selectivity of DMF was substantially decreased to 11%. Generally, an increase in temperature favoured hydrogenolysis regardless of the catalyst or other reaction parameters.[72] Thus, the temperature was further increased to 130 °C; as expected the reaction was very fast and complete conversion could be achieved within 10 min. Moreover, the calculation of reaction rate in terms of turnover frequency (TOF) was also very high of 3167 h^{-1} at 130 °C, but the selectivity of DMF was comparatively low of 66% despite higher TOF. Thus, a moderate temperature of 80 °C was used as an optimum temperature throughout the study.

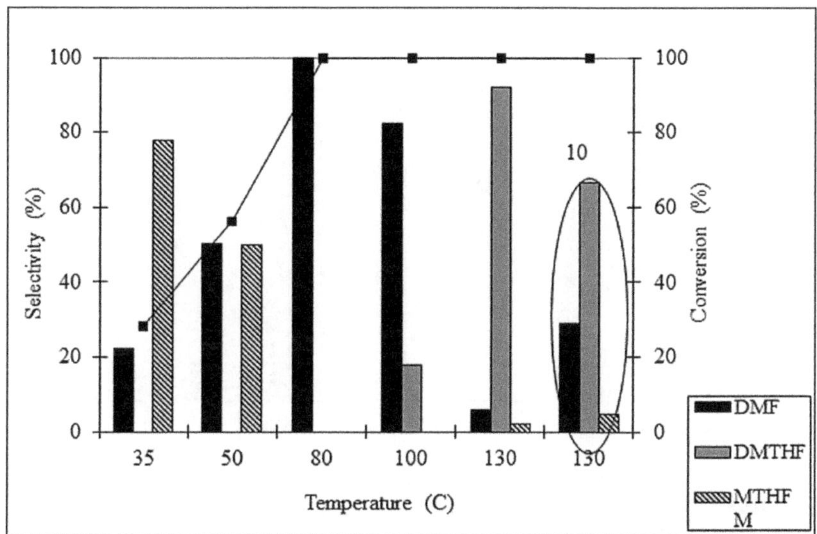

Figure 7.7 Effect of temperature on the conversion and product selectivity; reaction conditions: catalyst: substrate = 1 : 5; p_{CO_2} = 10 MPa, p_{H_2} = 1 MPa, reaction time = 2 h, H_2O = 1 mL; DMF = 2,5-dimethylfuran, DMTHF = 2,5-Dimethyltetrahydrofuran; MTHFM = tetrahydro-5-methyl-2-furanmethanol. The marked part shows the result of 10 min. Reproduced from ref. 87 with permission from The Royal Society of Chemistry.

In the present reaction system, 5-HMF was converted to DMF on Pd/C catalysts in $scCO_2$ medium. As presented in the catalyst screening section, in each case 1 mL of H_2O was used along with CO_2.

Figure 7.8 compares the results obtained in the presence and in the absence of H_2O, which shows that there was no effect on the conversion of HMF whether the reaction was conducted with or without H_2O. However, product components are strongly influenced by the addition of H_2O. For instance, MTHFM (selectivity = 95.7%) was the major product in $scCO_2$, whereas 95.3% of 5-methyl-2-furanyl methanol (MFM) and 4.7% of DMF constitute the product mixture in only a H_2O medium. Unfortunately, when the reaction was conducted in pure $scCO_2$, DMF remains undetected by GCMS analysis. Unlike $scCO_2$, DMF was detected with very low selectivity in H_2O. From these clues, H_2O emerged as a required element under the applied condition, thus, a combined effect of CO_2 and H_2O was studied. The mole ratio of $H_2O:CO_2$ was varied from 0.16 : 1 to 1.3 : 1 (Figure 7.8). As a result, the selectivity of DMF was changed and strikingly high for the $H_2O:CO_2$ ratio of 0.32 : 1. From Scheme 7.2 it can be suggested that the hydrogenation of the –CHO group followed by the hydrogenolysis of two –CH_2OH groups of the 5-HMF result in DMF. In the presence of a Pd catalyst, the hydrogenation of –CHO group could be easily occurred in the $scCO_2$ medium because of the large amount of dissolved hydrogen. However, utilization of H_2O promoted hydrogenolysis. In the presence of H_2O, the reaction medium becomes acidic with a pH ≅ 3.4 because of the formation of

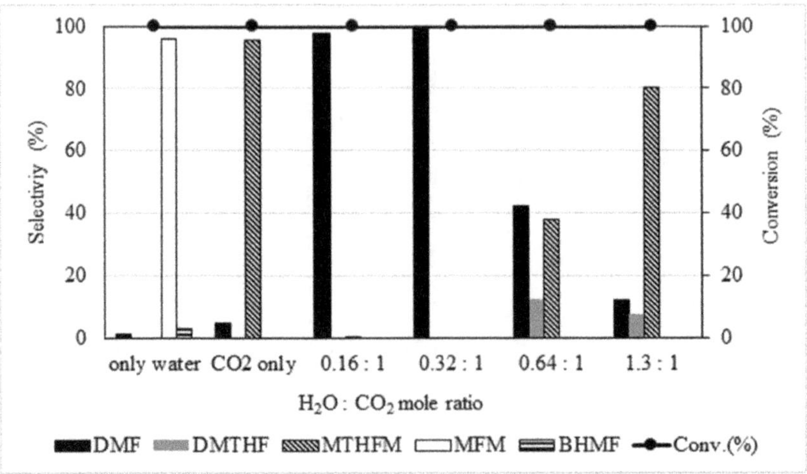

Figure 7.8 Variation of $H_2O:CO_2$ mole ratio; reaction conditions: catalyst: substrate = 1 : 5; p_{CO_2} = 10 MPa, p_{H_2} = 1 MPa, reaction time = 2 h, temperature = 80 °C; DMF = 2,5-dimethylfuran, DMTHF = 2,5-dimethyltetrahydrofuran; MTHFM = tetrahydro-5-methyl-2-furanmethanol, BHMF = 2,5-bis(hydroxymethyl)furan (MF), MFM = 5-methylfuranyl methanol. Reproduced from ref. 87 with permission from The Royal Society of Chemistry.

Table 7.2 Effect of different solvents along with water on the conversion and selectivity of DMF. Adapted from ref. 87 with permission from The Royal Society of Chemistry.[a]

Entry	Solvent	Conv. (%)	DMF selectivity (%)
1	Hexane	23.3	2.5
2	1-butanol	59.2	38.1
3	THF	40.2	26.2
4	$scCO_2$	100	100

[a]Reaction conditions: catalyst : substrate = 1 : 5; $p_{CO_2} = 10$ MPa; $p_{H_2} = 1$ MPa; reaction time = 2 h; temperature = 80 °C; H_2O = 1 mL; DMF = 2,5-dimethylfuran.

carbonic acid through the interaction with CO_2.[73] This mild acidic condition boosted the hydrogenolysis of CH_2OH groups, and abruptly enhanced the selectivity of DMF. However, an increased amount of H_2O decreased the selectivity of DMF. As mentioned before, MFM and DMF formed at the same time as a major and minor product, respectively, in only H_2O system. An excess of H_2O replicates the same scenario, but due to the presence of the CO_2 in the system MFM as well as DMF get extracted into the CO_2 phase and then hydrogenated to their corresponding products. Thus an optimized $H_2O : CO_2$ mole ratio of 0.32 : 1 was required to achieve highest selectivity of DMF under the present reaction condition.

The combined effect of $scCO_2$–H_2O was justified when the results were compared with the reaction performed in the different organic solvents–H_2O medium (Table 7.2). Applying the similar condition, the conversion and the selectivity of DMF was dropped significantly. Interestingly, lowest selectivity of DMF (2.5%) was detected in hexane; a non-polar solvent and comparable with $scCO_2$ in terms of dielectric constant (Entry 1). Only 1-butanol shows comparatively better selectivity of DMF (38%) (Entry 2). The results of the conversion of 5-HMF to DMF in organic solvent–H_2O mixture confirmed the positive effect of the acidic environment on the hydrogenolysis reaction to achieve very high selectivity of DMF.

7.3.3 Furfural to 2-Methylfuran (2-MF)

Like 5-HMF, furfural also considered as a promising platform molecule, which can be converted to 2-MF in a $scCO_2$–H_2O medium within the reaction time of 10 min representing a very fast reaction of the selective removal of oxygen from the side group. 2-MF is also used as a fuel additive with research octane number of 131 and has very low solubility in water.[21] It is challenging to produce 2-MF without any unnecessary hydrogenation of C=C of the furan ring. Table 7.3 represents the effect of hydrogen pressure and the reaction time on the product distribution of furfural hydrogenation. At a higher hydrogen pressure of 1 MPa and the longer reaction time of 2 h, furfural was completely converted to saturated 2-methyltetrahydrofuran

Table 7.3 Effect of different reaction parameters on the transformation of furfural in $scCO_2$. Adapted from ref. 87 with permission from The Royal Society of Chemistry.[a]

Entry	p_{H_2} (MPa)	Time (min)	Conv. (%)	Product selectivity (%)		
				Furfuryl alcohol	2-MF	2-methyltetrahydrofuran
1	1	120	100	0.0	0.0	100.0
2	1	20	100	0.0	74.1	24.9
3	1	10	100	0.0	100	0.0
4	0.2	5	13.0	91.9	8.1	0.0
5	0.2	10	21.1	0.0	100	0.0
6	0.2	15	65.5	0.0	59.9	40.1

[a]Reaction conditions: catalyst : substrate = 1 : 5, p_{CO_2} = 10 MPa, reaction time = 2 h, temperature = 80 °C, H_2O = 1 mL.

(Entry 1). However, over hydrogenation could be restricted by shortening the reaction time to 10 min and 2-MF was readily formed with 100% selectivity (Entry 3). In a time course of furfural hydrogenation at 0.2 MPa of hydrogen pressure, the conversion was changed from 13 to 65.5% as the time changes from 5 to 15 min. (Entries 4 to 6). Furfural was rapidly converted to furfuryl alcohol with 91.9% selectivity (Entry 4) within the reaction time of 5 min and after 10 min furfuryl alcohol was completely diminished followed by the enhanced selectivity of 2-MF (Entry 5). After extending the reaction time to 2 h, the furan ring was hydrogenated to form 2-methyltetrahydrofuran (Entry 6). Generally, the furan ring has a strong interaction with the surface of the metal catalyst through the *p*-bond of the ring and the d-orbital of the metal,[74,75] which results the hydrogenation of the furan ring depending on the surface hydrogen coverage.[76] Thus, it is possible to obtain DMF and 2-MF effectively in a simple and sustainable way from their respective precursor in $scCO_2$–H_2O medium without using any organic solvents.

7.3.4 THFA to 1,5-Pentanediol (1,5-PD)

As mentioned in the previous section, like 5-HMF, furfural is also blessed with a rich hydrogenation chemistry with the conversion of –CHO to –CH_2OH and hydrogenation of the furan ring.[21] THFA a saturated product of furfural is also a renewable feedstock can be converted to diol in the presence of hydrogen. Diols or polyols are –OH group containing compounds has a wide range of applications including the production of rigid and flexible type of polyurethane foams. 1,5-PD is one of the attractive monomers to find its application as an intermediate for the different chemical synthesis, ink and coatings, plasticizer as well as solvent-borne plastics.[77–79] The synthesis of 1,5-PD through a non-petroleum based route is challenging and increasing attention because of the increasing petroleum prices. Schniepp and Geller developed a three step method for conversion of THFA to 1,5-PD, which involves (i) dehydration of THFA to dihydropyran, (ii) hydration of dihydropyran to 5-hydroxypentanal and finally (iii) hydrogenation of

5-hydroxypentanal to 1,5-PD. However, this three steps reaction has a drawback in terms of the loss of the desired compound because of separation and purification, which subsequently resulted low yield of 1,5-PD.[77] Adkins and Connor attempted direct conversion of furfuryl alcohol to 1,5-PD using Cu-chromite catalyst, but they reached only 30% yield of the targeted 1,5-PD.[78] Koso *et al.* performed some outstanding work in this field and introduced Rh-ReO_x/SiO_2 catalyst to achieve a very high yield of 77% at 120 °C and 8 MPa of hydrogen pressure. They found that without any modification, Rh/SiO_2 produces 1,2-pentanediol (1,2-PD) as a final product with a very low conversion. To study the promotional effect, instead of Re, Mo and W were also applied along with Rh. In addition to Re, Mo also drastically improved the yield of 1,5-PD and predicted a synergic effect between Rh and the specific additive, which is very important for the hydrogenolysis of THFA. Addition of Re to Rh/SiO_2 improved the adsorption of THFA on catalyst through an interaction involving the –OH group, which influenced the selectivity.[79,80] Dumesic and co-workers also suggested the promotional effect of ReO_x on the hydrogenolysis of the C–O bond of polyols over Rh-ReO_x/C.[81] A direct approach to the generation of 1,5-PD from furfural was attempted by Xu *et al.* using Pt/$CoAlO_4$ catalyst in ethanol and they ended up with the formation of a mixture of THFA and 1,5-PD.[82]

Hence, there are several strategies to improve the transformation of THFA to 1,5-PD involving, multistep process, harsh reaction condition or using metal additive. To design a process of the transformation of THFA to 1,5-PD in a simple and convenient way is still challenging. For hydrogenolysis of THFA under the mild reaction condition, $scCO_2$ can be introduced as a reaction medium. Interest behind the use of $scCO_2$ is that the tuning of the solvent properties like density, diffusion rate and heat capacity, which might allow kinetic control of the reaction is possible. The hydrogenolysis of THFA was conducted using Rh and Pd catalysts supported on MCM-41, C and Al_2O_3. Table 7.4 shows that both the metals exhibited reasonable catalytic

Table 7.4 A comparison between Rh and Pd catalyst on the conversion of THFA to 1,5-PD. Adapted from ref. 86 with permission from The Royal Society of Chemistry.[a]

			Selectivity (%)		
Entry	Catalyst	Conv. (%)	1,5-PD	1,2-PD	Pentanol
1	Rh/MCM-41	80.5	91.2	0.0	8.8
2	Rh/Al_2O_3	60.0	61.9	0.0	30.9 (7.2)
3	Rh/C	34.8	49.8	2.3	7.9 (40.0)
4	Pd/MCM-41	50.5	12.6	77.4	10.0
5	Pd/Al_2O_3	32.5	9.1	0.9	47.5 (57.5)
6	Pd/C	48.9	99.2	0.3	0.5
7[b]	Rh/MCM-41	90.2	74.8	0.1	25.1

[a]Reaction conditions: catalyst = 0.1 g, THFA = 0.4 g; temperature = 80 °C; p_{CO_2} = 14 MPa, p_{H_2} = 4 MPa, time = 24 h; data in parenthesis represents by-products; 1,5-PD = 1,5-pentanediol, 1,2-PD = 1,2-pentanediol.

[b]Reaction time = 36 h.

activity, but no definite trend was detected depending on the support used. Applying the same reaction condition (temperature = 80 °C, p_{CO_2} = 14 MPa, p_{H_2} = 4 MPa, reaction time = 24 h) the conversion of THFA followed the order of MCM-41 > Al_2O_3 > C and C > MCM-41 > Al_2O_3 over Rh and Pd, respectively. Analysis of the product mixture revealed that 1,5-PD produced with the highest selectivity of 91.2% on Rh/MCM-41 (Entry 1) followed by Rh/Al_2O_3 (61.1%) and Rh/C (49.8%) (Entries 2 and 3). However, 1,2-PD and pentanol was observed as the primary product with the selectivity of 77.4% and 47.5% on Pd/MCM-41 and Pd/Al_2O_3, respectively (Entries 4 and 5). Moreover, very high selectivity of 1,5-PD (99.2%) was detected over Pd/C, however, the conversion was relatively low of 48% (Entry 6). Therefore, targeted on the high conversion and selectivity of 1,5-PD, Rh/MCM-41 was chosen as a potential candidate for further studies in the pressurized CO_2 medium.

Figure 7.9 represents the results of THFA conversion at different CO_2 pressure ranging from 0 to 18 MPa at 80 °C and the fixed hydrogen pressure of 4 MPa. Depending on the pressure, the conversion of THFA was increased from <10% (8 MPa) to 80% (14 MPa) and then remain constant without affecting the selectivity of 1,5-PD. This result might be attributed to the solubility of the substrate in $scCO_2$ (Figure 7.10(a)–(f)). At low pressure, the system consisted of a THFA-rich liquid phase and a gaseous phase. An increase in pressure resulted in an increase in the solubility of THFA in CO_2 and the system turned into a homogeneous THFA–CO_2–H_2 phase when the pressure reached at or above 12 MPa. It has to be mentioned that the temperature and hydrogen pressure were kept constant. Hence, the concentration of THFA in the CO_2–H_2-rich phase was steadily increased with

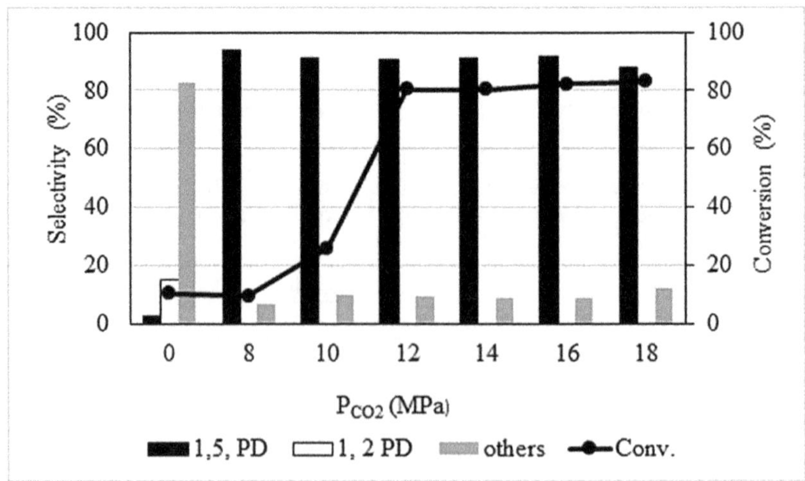

Figure 7.9 Effect of CO_2 pressure on the conversion and selectivity of THFA hydrogenolysis. Reaction conditions: catalyst = 0.1 g, THFA = 0.4 g; temperature = 80 °C; p_{H_2} = 4 MPa, reaction time = 24 h.
Reproduced from ref. 86 with permission from The Royal Society of Chemistry.

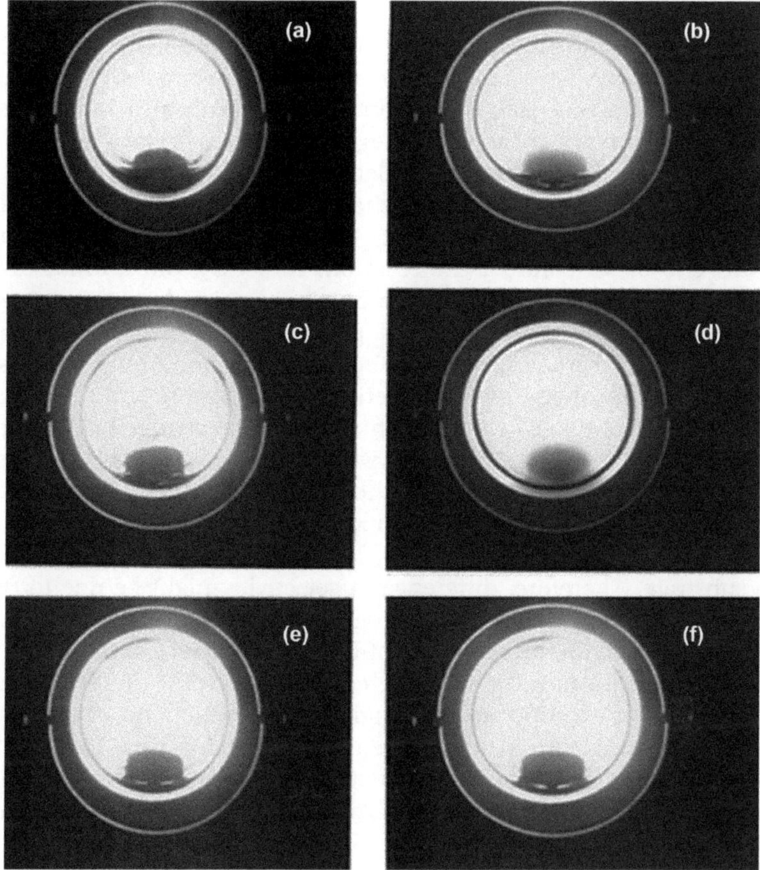

Figure 7.10 Phase behaviour of THFA in CO_2 at the fixed pressure of H_2 ($p_{H_2} = 4$ MPa). (a) only THFA and in the presence of CO_2 (b) 8, (c) 10, (d) cloud point, (e) 12 and (f) 14 MPa.
Reproduced from ref. 86 with permission from The Royal Society of Chemistry.

pressure and the maximum conversion was achieved in the homogeneous phase condition. In the absence of CO_2 ($p_{CO_2} = 0$ MPa; solvent-less), the conversion of THFA was low of 10% and the analysis of the product mixture failed to provide any evidence of the formation of 1,5-PD.

The Rh/MCM-41 catalyst was active for the reaction after the application of 4 MPa of hydrogen pressure. A critical role of hydrogen pressure revealed that at lower pressure (2 MPa) the yield of 1,5-PD was very low, but increase in pressure (>4 MPa) results higher conversion but the product yield dropped significantly.

Optimization of temperature for this reaction also improved the catalytic performance of Rh/MCM-41 in terms of the selectivity of 1,5-PD. Under the fixed pressure condition, the temperature of the system varied from

60 to 120 °C. The Rh/MCM-41 catalyst was inactive at lower temperature (<60 °C) and the complete conversion of THFA was observed at 120 °C. The increased conversion with temperature may be attributed to the dramatic increase of the reaction rate in the liquid phase and also in the gaseous phase, which overcompensates the negative effect of phase separation at higher temperature.[83] However, the selectivity of 1,5-PD was decreased from 91% to 46% as the temperature was changed from 80 °C to 120 °C despite complete conversion of THFA at 120 °C. Hence, an optimum temperature of 80 °C was chosen to conduct the reaction. In $scCO_2$, one cannot expect a straight forward effect of temperature on the reaction because the change in temperature affects the density of the medium and consequently the solubility of the substrate. Under a fixed pressure condition, the density of the medium was changed with temperature. For instance, at 14 MPa, the density changes from 0.384 to 0.256 $g\,mL^{-1}$ with the change in the temperature from 80 to 120 °C (NIST chemistry web book), respectively. As a result, there was a possibility of phase separation, thus, single phase might be split into two phases because at higher temperature the CO_2 phase becomes less dense. Hence, at lower densities, $scCO_2$ are more gas-like and poorer solvent for the substrate. Therefore, difference in selectivity might be due to the difference in the reaction rate of the hydrogenolysis of THFA in various phases.

From a time-dependent reaction profile very slow transformation of THFA could be predicted under the working condition as <5% of THFA was converted after 2 h of reaction. Interestingly, except 1,5-PD no other products were within the detection limit. Shortened the reaction time to <2 h, THFA remain unconverted. Thus, it is difficult to determine the intermediate formed under the applied condition. Evidently, at 80 °C the hydrogenolysis of THFA required at least 6 h to achieve ~40% of conversion, an extension of reaction time to 36 h, conversion was slightly improved from 80 to 90%, but the selectivity of 1,5-PD started to decrease (Entry 7).

The transformation of THFA to 1,5-PD is a challenging process and an appreciable catalytic performance was observed in $scCO_2$ under a mild reaction condition. Identification of the reaction intermediate in $scCO_2$ will be helpful to for catalyst designing as well as to improve the reaction strategy to achieve the highest yield of the targeted compound in a clean and green way.

7.4 Conclusion

The combination of a heterogeneous catalyst and $scCO_2$ offers innovative and unexplored opportunities to establish novel transformation pathways of biomass-derived compounds. Several potential scenarios on the usage of biomass can be foreseen. One of the greatest hopes is the preparation of biofuels in a sustainable (in terms of economic and environmental) way.

This chapter has highlighted the development of different strategies to convert 5-HMF and furfural into fuel and non-fuel related compounds. 5-HMF could be transformed into linear alkanes *via* hydrogenation/ hydrogenolysis and dehydration under a mild reaction conditions.

Hydrogenation, followed by the hydrogenolysis, of 5-HMF and furfural produces DMF and 2-MF, respectively, with very high yields in $scCO_2$–H_2O mixture. Due to the unique tuneable properties of $scCO_2$, it was possible to produce three different value-added products from HMF simply by changing the CO_2 pressure in the presence of H_2O. $ScCO_2$ was exploited in the transformation of THFA, to diol (precursor of monomer). THFA was successfully converted to 1,5-PD with high yield under a mild reaction condition.

These exciting results on the use of $scCO_2$ and $scCO_2$–H_2O offer several interesting prospects related to the preparation of different valuable chemicals from biomass. For the future, the challenge is the development of energy efficient large-scale technologies to produce biofuels and fine chemicals through the intelligent use of biomass that could improve environmental and economic sustainability.

Acknowledgements

This work is partly funded by the New Energy and Industrial Technology Development (NEDO) Japan.

References

1. D. C. Elliott, in *The Encyclopedia of Energy*, ed. C. J. Cleveland, Elsevier, Oxford, United Kingdom, 2005, vol. 1, pp. 163–174.
2. A. C. O'Sullivan, *Cellulose*, 1997, **4**, 173–207.
3. Y. Nishiyama, P. Langan and H. Chanzy, *J. Am. Chem. Soc.*, 2002, **124**, 9074–9082.
4. A. Brandt, J. Grasvik, J. P. Hallett and T. Welton, *Green Chem.*, 2013, **15**, 550–583.
5. E. Hirst, *Pure Appl. Chem*, 1962, **5**, 53–66.
6. C. K. Boyce, M. A. Zwieniecki, G. D. Cody, C. Jacobsen, S. Wirick, A. H. Knoll and N. M. Holbrook, *Proc. Natl. Acad. Sci. U. S. A.*, 2004, **101**, 17555–17558.
7. N. H. Bhuiyan, G. Selvaraj, Y. Wei and J. King, *Plant Signaling Behav.*, 2009, **4**, 158–159.
8. S. J. Liljegren, G. S. Ditta, Y. Eshed, B. Savidge, J. L. Bowman and M. F. Yanofsky, *Nature*, 2000, **404**, 766–770.
9. K. V. Sarkanen and C. H. Ludwig, *Lignins: Occurrence, Formation, Structure and Reactions*, Wiley-Interscience, New York, NY, USA, 1971.
10. K. Freudenberg and A. C. Neish, *Constitution and Biosynthesis of Lignin.*, 1968.
11. J. N. Chheda, G. W. Huber and J. A. Dumesic, *Angew. Chem., Int. Ed.*, 2007, **46**, 7164–7183.
12. H.-J. Huang, S. Ramaswamy, U. Tschirner and B. Ramarao, *Sep. Purif. Technol.*, 2008, **62**, 1–21.

13. T. Werpy and G. R. Petersen, *Top Value Added Chemicals From Biomass - Volume I: Results of Screening for Potential Candidates from Sugars and Synthesis Gas* the Pacific Northwest National Laboratory (PNNL) and the National Renewable Energy Laboratory (NREL), 2004.
14. J. Lewkowski, *ARKIVOC*, 2001, **i**, 17–54.
15. C. Moreau, M. N. Belgacem and A. Gandini, *Top. Catal.*, 2004, **27**, 11–30.
16. X. L. Tong, Y. Ma and Y. D. Li, *Appl. Catal., A*, 2010, **385**, 1–13.
17. C.-H. Zhou, X. Xia, C.-X. Lin, D.-S. Tong and J. Beltramini, *Chem. Soc. Rev.*, 2011, **40**, 5588–5617.
18. A. A. Rosatella, S. P. Simeonov, R. F. Frade and C. A. Afonso, *Green Chem.*, 2011, **13**, 754–793.
19. M. E. Zakrzewska, E. Bogel-Lukasik and R. Bogel-Lukasik, *Chem. Rev.*, 2011, **111**, 397–417.
20. R.-J. van Putten, J. C. van der Waal, E. de Jong, C. B. Rasrendra, H. J. Heeres and J. G. de Vries, *Chem. Rev.*, 2013, **113**, 1499–1597.
21. J. P. Lange, E. van der Heide, J. van Buijtenen and R. Price, *ChemSusChem*, 2012, **5**, 150–166.
22. M. Schlaf, *Dalton Trans.*, 2006, 4645–4653.
23. R. A. Sheldon, *Green Chem.*, 2007, **9**, 1273–1283.
24. P. T. Anastas and J. C. Warner, *Green Chemistry: Theory and Practice*, Oxford University Press, New York, 1998.
25. J. H. Clark and S. J. Tavener, *Org. Process Res. Dev.*, 2007, **11**, 149–155.
26. A. Fürstner, L. Ackermann, K. Beck, H. Hori, D. Koch, K. Langemann, M. Liebl, C. Six and W. Leitner, *J. Am. Chem. Soc.*, 2001, **123**, 9000–9006.
27. K. Wittmann, W. Wisniewski, R. Mynott, W. Leitner, C. L. Kranemann, T. Rische, P. Eilbracht, S. Kluwer, J. M. Ernsting and C. J. Elsevier, *Chem. – Eur. J.*, 2001, **7**, 4584–4589.
28. M. Chatterjee, A. Chatterjee, P. Raveendran and Y. Ikushima, *Green Chem.*, 2006, **8**, 445–449.
29. M. Chatterjee, A. Chatterjee, H. Kawanami, T. Ishizaka, T. Suzuki and A. Suzuki, *Adv. Synth. Catal.*, 2012, **354**, 2009–2018.
30. M. Chatterjee, T. Ishizaka, A. Suzuki and H. Kawanami, *Chem. Commun.*, 2013, **49**, 4567–4569.
31. M. Chatterjee, T. Ishizaka and H. Kawanami, *Green Chem.*, 2014, **16**, 4734–4739.
32. L. Reynolds, J. Gardecki, S. Frankland, M. Horng and M. Maroncelli, *J. Phys. Chem.*, 1996, **100**, 10337–10354.
33. S. G. Kazarian, M. F. Vincent, F. V. Bright, C. L. Liotta and C. A. Eckert, *J. Am. Chem. Soc.*, 1996, **118**, 1729–1736.
34. D. C. Rideout and R. Breslow, *J. Am. Chem. Soc.*, 1980, **102**, 7816–7817.
35. H. C. Hailes, *Org. Process Res. Dev.*, 2007, **11**, 114–120.
36. S. M. Cenci, L. R. Cox and G. A. Leeke, *ACS Sustainable Chem. Eng.*, 2014, **2**, 1280–1288.
37. B. Bhanage, *Chem. Commun.*, 1999, 1277–1278.

38. J. Beck, J. Vartuli, W. J. Roth, M. Leonowicz, C. Kresge, K. Schmitt, C. Chu, D. H. Olson and E. Sheppard, *J. Am. Chem. Soc.*, 1992, **114**, 10834–10843.
39. M. Grün, K. K. Unger, A. Matsumoto and K. Tsutsumi, *Microporous Mesoporous Mater.*, 1999, **27**, 207–216.
40. R. Mokaya and W. Jones, *Phys. Chem. Chem. Phys.*, 1999, **1**, 207–213.
41. R. Mulukutla, *Phys. Chem. Chem. Phys.*, 1999, **1**, 2027–2032.
42. D. L. Klass, *Biomass for Renewable Energy, Fuels, and Chemicals*, Academic press, 1998.
43. D. Elliott, D. Beckman, A. Bridgwater, J. Diebold, S. Gevert and Y. Solantausta, *Energ. Fuel.*, 1991, **5**, 399–410.
44. R. Davda, J. Shabaker, G. Huber, R. Cortright and J. A. Dumesic, *Appl. Catal., B*, 2005, **56**, 171–186.
45. G. W. Huber, R. D. Cortright and J. A. Dumesic, *Angew. Chem., Int. Ed.*, 2004, **43**, 1549–1551.
46. G. W. Huber, J. N. Chheda, C. J. Barrett and J. A. Dumesic, *Science*, 2005, **308**, 1446–1450.
47. J. N. Chheda and J. A. Dumesic, *Catal. Today*, 2007, **123**, 59–70.
48. G. W. Huber and J. A. Dumesic, *Catal. Today*, 2006, **111**, 119–132.
49. H. Chen and W. Qiu, *Biotechnol. Adv.*, 2010, **28**, 556–562.
50. P. Alvira, E. Tomas-Pejo, M. Ballesteros and M. J. Negro, *Bioresour. Technol.*, 2010, **101**, 4851–4861.
51. Y. Roman-Leshkov, C. J. Barrett, Z. Y. Liu and J. A. Dumesic, *Nature*, 2007, **447**, U982–U985.
52. S. Zhong, R. Daniel, H. Xu, J. Zhang, D. Turner, M. L. Wyszynski and P. Richards, *Energy Fuel.*, 2010, **24**, 2891–2899.
53. J. B. Binder and R. T. Raines, *J. Am. Chem. Soc.*, 2009, **131**, 1979–1985.
54. M. Chidambaram and A. T. Bell, *Green Chem.*, 2010, **12**, 1253–1262.
55. S. Morikawa, *Noguchi Kenkyusho Jiho*, 1980, **23**, 39–44.
56. J. Jae, W. Zheng, R. F. Lobo and D. G. Vlachos, *ChemSusChem*, 2013, **6**, 1158–1162.
57. T. S. Hansen, K. Barta, P. T. Anastas, P. C. Ford and A. Riisager, *Green Chem.*, 2012, **14**, 2457–2461.
58. G. Brieger and T. J. Nestrick, *Chem. Rev.*, 1974, **74**, 567–580.
59. R. C. Mebane and A. J. Mansfield, *Synth. Commun.*, 2005, **35**, 3083–3086.
60. F. Alonso, P. Riente and M. Yus, *Acc. Chem. Res.*, 2011, **44**, 379–391.
61. F. Alonso, P. Riente and M. Yus, *Tetrahedron*, 2008, **64**, 1847–1852.
62. G. P. Boldrini, D. Savoia, E. Tagliavini, C. Trombini and A. Umani-Ronchi, *J. Org. Chem.*, 1985, **50**, 3082–3086.
63. M. Kidwai, V. Bansal, A. Saxena, R. Shankar and S. Mozumdar, *Tetrahedron Lett.*, 2006, **47**, 4161–4165.
64. T. Thananatthanachon and T. B. Rauchfuss, *Angew. Chem.*, 2010, **122**, 6766–6768.
65. G.-H. Wang, J. Hilgert, F. H. Richter, F. Wang, H.-J. Bongard, B. Spliethoff, C. Weidenthaler and F. Schüth, *Nat. Mater.*, 2014, **13**, 293–300.

66. G. C. Luijkx, N. P. Huck, F. van Rantwijk, L. Maat and H. van Bekkum, *Heterocycles*, 2009, **77**, 1037–1044.
67. Y. Kwon, E. de Jong, S. Raoufmoghaddam and M. Koper, *ChemSusChem*, 2013, **6**, 1659–1667.
68. G. Bottari, A. J. Kumalaputri, K. K. Krawczyk, B. L. Feringa, H. J. Heeres and K. Barta, *ChemSusChem*, 2015, **8**, 1323–1327.
69. B. Saha, C. M. Bohn and M. M. Abu-Omar, *ChemSusChem*, 2014, **7**, 3095–3101.
70. J. Zhang, L. Lin and S. Liu, *Energy Fuel.*, 2012, **26**, 4560–4567.
71. S. Wang, V. Vorotnikov and D. G. Vlachos, *Green Chem.*, 2014, **16**, 736–747.
72. P. Rylander, *Catalytic Hydrogenation Over Platinum Metals*, Elsevier, 2012.
73. C. Roosen, M. Ansorge-Schumacher, T. Mang, W. Leitner and L. Greiner, *Green Chem.*, 2007, **9**, 455–458.
74. M. K. Bradley, J. Robinson and D. Woodruff, *Surf. Sci.*, 2010, **604**, 920–925.
75. C. J. Kliewer, C. Aliaga, M. Bieri, W. Huang, C.-K. Tsung, J. B. Wood, K. Komvopoulos and G. A. Somorjai, *J. Am. Chem. Soc.*, 2010, **132**, 13088–13095.
76. S. Wang, V. Vorotnikov and D. G. Vlachos, *ACS Catal.*, 2014, **5**, 104–112.
77. L. Schniepp and H. Geller, *J. Am. Chem. Soc.*, 1946, **68**, 1646–1648.
78. H. Adkins and R. Connor, *J. Am. Chem. Soc.*, 1931, **53**, 1091–1095.
79. S. Koso, I. Furikado, A. Shimao, T. Miyazawa, K. Kunimori and K. Tomishige, *Chem. Commun.*, 2009, 2035–2037.
80. S. Koso, N. Ueda, Y. Shinmi, K. Okumura, T. Kizuka and K. Tomishige, *J. Catal.*, 2009, **267**, 89–92.
81. M. Chia, Y. J. Pagan-Torres, D. Hibbitts, Q. Tan, H. N. Pham, A. K. Datye, M. Neurock, R. J. Davis and J. A. Dumesic, *J. Am. Chem. Soc.*, 2011, **133**, 12675–12689.
82. W. Xu, H. Wang, X. Liu, J. Ren, Y. Wang and G. Lu, *Chem. Commun.*, 2011, **47**, 3924–3926.
83. R. Tschan, R. Wandeler, M. S. Schneider, M. M. Schubert and A. Baiker, *J. Catal.*, 2001, **204**, 219–229.
84. P. Markewitz, W. Kuckshinrichs, W. Leitner, J. Linssen, P. Zapp, R. Bongartz, A. Schreiber and T. E. Müller, *Energy Environ. Sci.*, 2012, **5**, 7281–7305.
85. M. Chatterjee, K. Matsushima, Y. Ikushima, M. Sato, T. Yokoyama, H. Kawanami and T. Suzuki, *Green Chem.*, 2010, **12**, 779–782.
86. M. Chatterjee, H. Kawanami, T. Ishizaka, M. Sato, T. Suzuki and A. Suzuki, *Catal. Sci. Technol.*, 2011, **1**, 1466–1471.
87. M. Chatterjee, T. Ishizaka and H. Kawanami, *Green Chem.*, 2014, **16**, 1543–1551.

CHAPTER 8

Anti-solvent Effect of High-pressure CO_2 in Natural Polymers

ARTURO ÁLVAREZ-BAUTISTA[a,b] AND ANA MATIAS*[a,b]

[a] iBET, Instituto de Biologia Experimental e Tecnológica, Apartado 12, 2780-901 Oeiras, Portugal; [b] Instituto de Tecnologia Química e Biológica António Xavier, Universidade Nova de Lisboa, Av. da República, 2780-157 Oeiras, Portugal
*Email: amatias@ibet.pt

8.1 Introduction

Biopolymers are polymers or copolymers produced from the raw materials of renewable sources (biomass), such as corn, sugarcane, pulp mills and marine crustaceans. Renewable sources are well known to possess a shorter life cycle when compared to fossil fuels, which take thousands of years to form. Some environmental and socio-economic factors that are related to the growing interest in biopolymers are: great environmental impacts caused by the extraction and refining processes used to produce polymers from oil, the shortage of oil and the increase of petroleum price.

Because of this, and the versatility and wide application of natural polymers, the production of these materials has increased in recent years. However, the recovery of these materials is necessary after their usage to protect and conserve the environment. Thus, the recovery of biopolymers is presented as an advantageous alternative. In addition these processes must provide economic benefits to be considered optimal. Therefore they must

Green Chemistry Series No. 48
High Pressure Technologies in Biomass Conversion
Edited by Rafał M. Łukasik

Published by the Royal Society of Chemistry, www.rsc.org

meet certain aspects such as the reduction of raw material, the increase of free volume in landfills, reduction of resources, energy consumption and environmental and social benefits.

Recently, the production of valuable products from the different kinds of biomass has been investigated from lignocellulosic or marine biomass within a biorefinery concept. However, we are still lacking efficient and optimized separation processes for biomass-based polymers in biorefineries. Cellulose, hemicellulose and chitosan are some of the principal biopolymers that can be found in lignocellulosic and marine biomass, respectively. However, the production of such valuable compounds improving the overall techno-economic feasibility of biorefineries is thus still a big challenge.

Techniques employing a supercritical anti-solvent effect (SAS) have been proved to be the most appropriate technology[1,2] for the precipitation of biopolymers even leading to obtain particles of defined morphology. In order to get the best results it is necessary to know the thermodynamic and phase behaviour of the materials to choose the optimal process conditions. CO_2 has been widely used as anti-solvent for recovery of polymers and biopolymers due to its excellent properties, mild supercritical conditions and the immiscibility of polymers and biopolymers into the gas. This technique presents several advantages over the conventional methods, use of organic solvents is reduced, prevents thermal and mechanical solute degradation and solvent residual concentration is decreased.

Beside all these reasons and due to the versatility and wide application of natural polymers, the production volume of these materials has increased in recent years. However, the recovery of these materials is necessary after use to protect and conserve the environment. Thus, the recovery of biopolymers is presented as an advantageous alternative. This alternative, in turn, must provide economic benefits to be considered optimal. For it must meet certain aspects such as: the reduction of raw material, the increase of free volume in landfills, reduction of resources, energy consumption and environmental and social benefits.

8.2 Biopolymers

8.2.1 Cellulose

Cellulose is a polydisperse polymer with a degree of polymerization (DP) between 3500 and 37 000. The native cellulose is widely distributed in nature. The structure of cellulose is formed by the union of β-glucose molecules *via* β-1,4-glycosidic links, which makes it insoluble in water. Cellulose has a linear or fibrous structure, (Scheme 8.1) wherein multiple hydrogen bonds are established between hydroxyl groups of different juxtaposed glucose chains, making them highly resistant and insoluble in water. Cellulose is the main component of kapok, cotton, flax, hemp, ramie and wood. Currently, the cellulose is extracted from wood fibres, separating from other natural components such as lignin and hemicellulose.

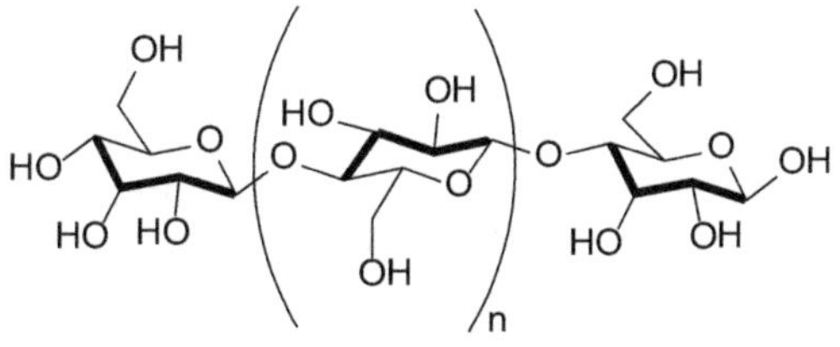

Scheme 8.1 The chemical structure of cellulose.

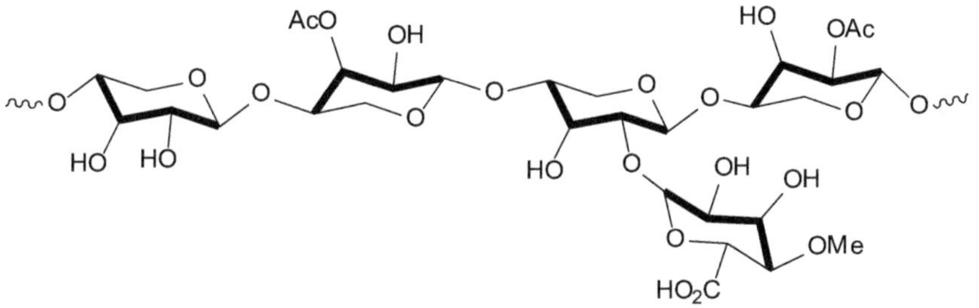

Scheme 8.2 The chemical structure of hemicellulose.

The regenerated cellulose is produced by precipitation from solutions of native cellulose in solvents in which it is insoluble. Cellulose makes up more than one-third of the vegetal matter. It is the most abundant organic compound in the world. There are several subtypes of cellulose. High molecular weight native cellulose, which is insoluble in an aqueous solution of 18% NaOH is called α-cellulose. The fraction which is soluble in the above solution, but insoluble in a solution of 8% is known as β-cellulose and the fraction soluble in a solution of 8% is called γ-cellulose.

The development of environmentally friendly low-cost solvents for cellulose is essential for the successful utilization of the cellulose as a component of polymeric materials. Normally the regenerated cellulose can be prepared directly through a dissolution, formation and regeneration process. This is an environmentally friendly process since most of the chemicals used can be recycled and reused. Modifying the regeneration parameters cellulose in different forms can be obtained: powders, fibres, spheres, hydrogels *etc.* Especially, the dissolution and regeneration is a "green", clean and physical process when "green" solvents are applied cellulose, leading to sustainable development, environmental preservation and energy conservations. Again, ionic liquids are presented as green solvents for cellulose.[3]

8.2.2 Hemicellulose

Hemicelluloses are made up of polymers of anhydrous sugar units joined by glycosidic linkages, formed by more than one type of sugar (hexoses or pentoses), and have ramifications and substitutions. Its role is to provide the bond between lignin and cellulose (Scheme 8.2). In natural state exists in amorphous form with a degree of polymerization not exceeding 200 (DP).

There are two types of hemicellulose: xylan and glucomannan. The use of hemicellulose is a great challenge today, as it represents the second most abundant biopolymer in the plant kingdom. Because of their large estates, hemicelluloses have a high potential for application. In conventional sulfite pulping processes, a large percentage of the hemicellulose fraction becomes degraded to oligomers or monomers. Hemicellulose in its native form is insoluble in water and to achieve dissolution of small amounts is necessary to use additives or lower pH values.

8.2.3 Chitosan

Chitosan is a polysaccharide obtained by deacetylating chitin, which is the major constituent of the exoskeleton of crustaceous water animals. Chemically, chitosan corresponds to (1-4)-2-amino-2-deoxy-β-D-glucan and includes a family of linear heteropolysaccharides composed of units of β-(1-4)-D-glucosamine (deacetylated units GlcN) and *N*-acetyl-D-glucosamine (acetylated units, GlcNAc) (Scheme 8.3). Furthermore, the distribution of these GlcNAc units along the chains of chitosan, which varies with the preparation method, influences the solubility of the polymer, and in intra/inter molecular interactions. Commercial chitosan is mainly produced by deacetylating this chitin material. The quality and properties of chitosan products may vary widely because many factors in the manufacturing process can influence the characteristics of the final product. Chitosan is

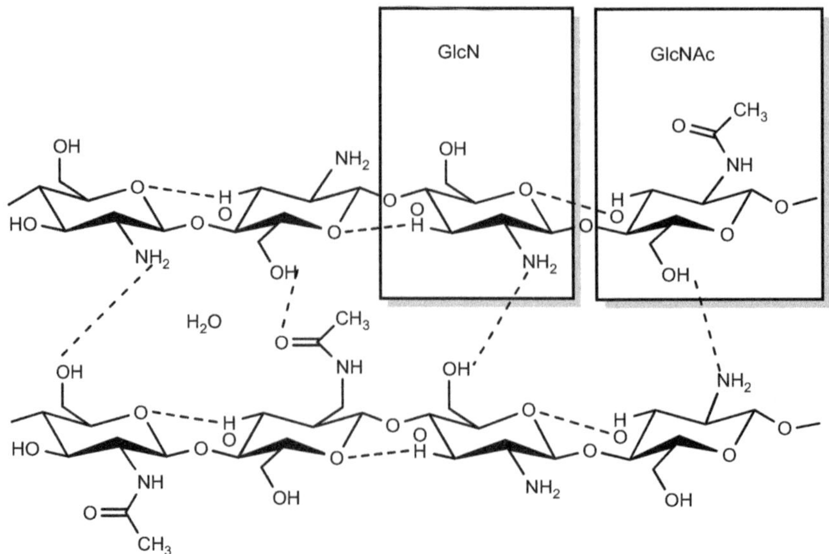

Scheme 8.3 Intra- and inter-molecular interactions of chitosan.

similar to cellulose in being a polysaccharide, but the repeating unit of the polymer backbone is not glucose but glucosamine, the amino group of which is largely unacetylated.

Due to the large use of chitosan in different industries, including biomedical applications as hydrogels,[4] this biopolymer production has increased in the last decade. For this reason, the search for different methods of recovery of biopolymers employing environmental mild conditions to avoid counterproductive damage is necessary.

Unfortunately chitosan is insoluble in a large amount of conventional organic solvents. This limits its applications. Recently ionic liquids (ILs), which have a great solvating potential, have been regarded[5] as a very efficient substitute for conventional solvents in dissolving this biopolymer.[6]

8.3 Anti-solvent Effect

8.3.1 Anti-solvent Effect to Regenerate Biopolymers

Biopolymers such as cellulose, hemicellulose or chitosan have strong inter- and intra-molecular hydrogen bonds, making them very hard to dissolve in common organic solvents. In this case, the dissolution efficiency of a solvent is determined by the ability of disrupting the hydrogen bonds. Some solvent systems have already been developed to dissolve biopolymers such as cellulose and hemicellulose. Some examples comprise thiourea–H_2O; NaOH–urea; LiOH–urea–H_2O; molten salt hydrates ($LiSCN \cdot 2H_2O$) or other systems like dimethylsulfoxide (DMSO) and tetrabutyl-ammonium fluoride (TBAF).[7] More recently, the ability of ILs to dissolve natural polymers has attracted great attention. Numerous studies have been published demonstrating the great potential of ILs as efficient and alternative solvents for biopolymers dissolution.[7–10] The use of ILs can overcome some drawbacks of the conventional solvents, such as high thermal stability and negligible vapour pressure.[10–13] However, the separation of the biopolymer and the solvent, ILs or others, is always a challenge[9] being extremely important to develop effective strategies to separate the biopolymers from the solvent and, in an environmental-friendly perspective, recycle a high percentage of the used solvent.

The regeneration of a biopolymer from biomass always involves first a dissolution step, where the biomass is dissolved in an adequate solvent (*e.g.* DMSO or ILs),[8,14] followed by a second step involving the precipitation of the different part of the dissolved biomass by the addition of a precipitating solvent by the called anti-solvent effect[10,15] (Figure 8.1).

The adequate recovery of different fractions from the biomass, in particular the regeneration of biopolymers from biomass is dependent on the dissolution solvent, type of biomass, pre-treatments applied and conditions used during the anti-solvent process.[10] Up to date, solvents such as water and ethanol, have been more commonly used as anti-solvents in the regeneration of biopolymers.[10] Nevertheless, in the frame of the biorefinery

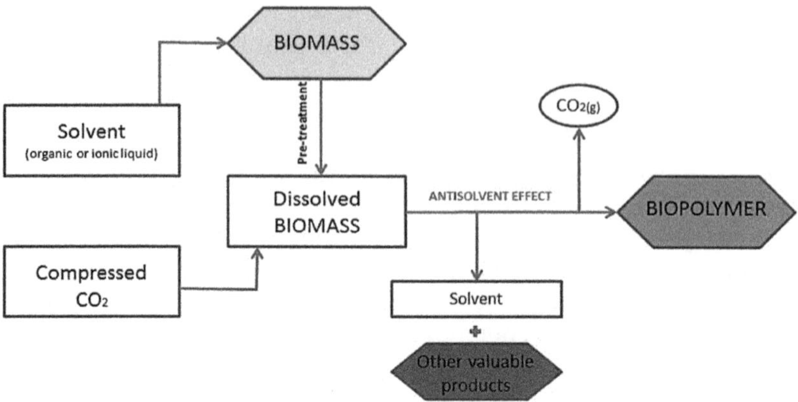

Figure 8.1 Schematic representation of the anti-solvent process for the regeneration of biopolymers from molecular solvents or ILs-based solutions.

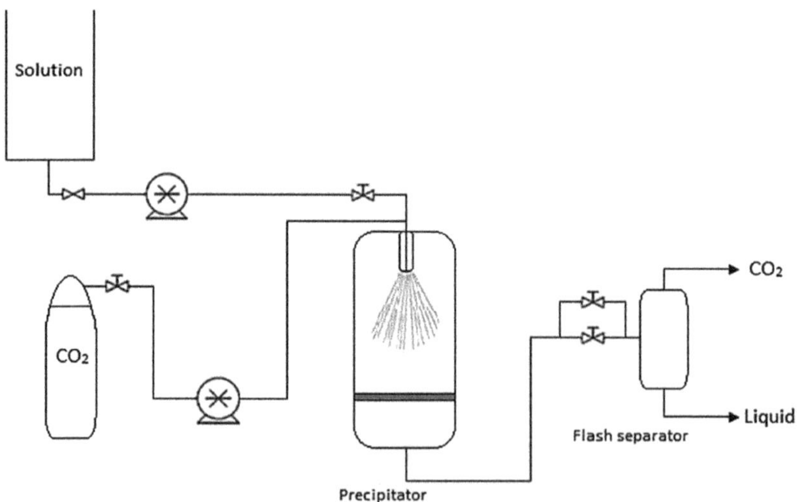

Figure 8.2 Schematic representation of the high pressure anti-solvent process.

concept, the use of these molecular solvents leads to other difficulties. The separation between solvent and anti-solvent needs to be solved in order to enable the development of a more environmentally friendly and highly efficient method for the regeneration of biopolymers from biomass.[9]

One of the effective processes to precipitate and dry polymers from organic solutions is driven by the addition of a gaseous anti-solvent to a polymer solution causing the supersaturation of the polymer. The supersaturation effect is more evident especially under supercritical conditions (Supercritical Anti-solvent (SAS)). In the SAS process (Figure 8.2), the polymer is dissolved in a classical organic solvent and precipitated in contact with a dense gas as a result of the supersaturation effect[16] (Figure 8.2).

In the last two decades, techniques employing dense gases have emerged and have demonstrated to be an appropriate and alternative technology for precipitation of polymers and biopolymers, overcoming some limitations related to the use of classical methodologies like spray drying, recrystallization or liquid anti-solvent. Exploring the particular properties of dense fluids, (*e.g.* the tuning of the fluid solvent power with minor changes in pressure and temperature) drawbacks of the conventional precipitation processes like the use of high shear forces, high amounts of classical solvents, high temperature and the presence of solvent residual concentration on the final product, can be overcome.[16–19]

8.3.2 Anti-solvent Effect of Compressed CO_2

Carbon dioxide, the most widely used dense gas, has been used as anti-solvent to assist the recovery of polymers and biopolymers from biomass. The process, which takes advantage of the fact that biopolymers are soluble in organic solvents but not in CO_2, can be carried out at mild temperatures (due to CO_2 low critical temperature) avoiding thermal degradation of labile compounds.[16,18,19]

Furthermore, CO_2 is non-flammable, abundant and relatively low toxic and can be easily recycled which are very important specifications from the industrial point of view. Due to its fast separation from the final products during the depressurization, the post processing is easier in comparison with the traditional methods, making CO_2 the elected solvent for processing products for human consumption.[8,16,20]

To date, only five pieces of work have been published reporting the use of CO_2 and compressed CO_2 as an anti-solvent to regenerate biopolymers from molecular solvents or IL-solutions. The type of operation, solvent of biopolymer, processing conditions and yield of recovery are depicted in Table 8.1.

In 2010, Haimer and co-workers, developed an anti-solvent CO_2-assisted precipitation process to prepare spherical hemicellulose microparticles from DMSO[14] and DMSO : water mixtures (Table 8.1). They performed batch and semi-continuous experiments to study the hemicelluloses solubility in DMSO : CO_2 and DMSO : CO_2 : water mixtures up to 15 MPa. Combining the Mukhopadhyay and Dalvi models[21] and Andreatta *et al.*[22] data for the binary DMSO : CO_2 system, they observed that CO_2 anti-solvent precipitation of hemicelluloses at 40 °C was achieved already in the pressure range of 6–9 MPa. In particular at 40 °C, beyond the critical pressure of the binary mixture DMSO : CO_2 (8.7 MPa) the results revealed that almost the total amount of hemicelluloses had precipitated. The method proposed by the authors showed to be adequate to prepare hemicellulose microparticles adjusted in size and morphology by the use of water in the DMSO : CO_2 mixture. Haimer's results also demonstrated that neither the dissolution temperature of hemicelluloses in the organic solvent nor the amount of dissolved

Table 8.1 The use of CO_2 and compressed CO_2 as an anti-solvent to regenerate biopolymers from molecular solvents or IL-solutions.

Regenerated biomass				Anti-solvent conditions					
Type	Recovery (%, w/w)	DP	Solvent	p (MPa)	T (°C)	t (h)	Operation type	Observations	Ref.
Hemicellulose	n.a.	n.a.	DMSO and DMSO–H_2O	15	40	n.a.	batch & semi-continuous	Purification of hemicelluloses from lignin was a side-effect of $scCO_2$ anti-solvent precipitation.	14
Chitin	95 (after 5×1 h batch cycle)	n.a.	[emim][OAc]	7.6–10.3	35–40	1; 2; 4	batch		23
Chitosan	22–62	—	[bmim][OAc]	6–12	25	1–10	batch	9.0 wt% chitosan/[bmim][OAc]	8, 9
	62	900[a] 1600[b] 600[c]		6.6	25	5			
	62	—		5.4	15	5			
Cellulose (microcrystalline cellulose MCC)	60	202[a] 282[b] 136[c]	[bmim][OAc]	6.6	25	3	batch	Influence of a co-solvent (DMSO; DMI and DMF) addition was studied	8
	65	157–175	[bmim][OAc]: DMSO	6.6–20	25	3		DMI and DMF demonstrated to have similar effects to DMSO	
Cellulose	60	n.a.	[bmim][OAc] solution	6–18	25	3	batch	Comparison with water, ethanol and acetonitrile as anti-solvent (yields >90%)	7

[a]Anti-solvent compressed CO_2.
[b]Native biopolymer.
[c]Anti-solvent ethanol, n.a. = not available.

hemicellulose have influence on particles morphology and size. Later, Barber and co-workers[23] demonstrated that CO_2 (gaseous or even supercritical) could be used as a coagulation solvent for biopolymer–IL solutions. This work encouraged Sun and co-workers in 2014[8,9] to study the regeneration of chitosan and cellulose from acetate ILs-solutions using compressed CO_2. Using a 9.0 wt% chitosan/1-butyl-3-methylimidazolium acetate ([bmim][OAc]) solution, regenerated chitosan was obtained from a pressure of 6.6 MPa at 25 °C. The regenerated chitosan by compressed CO_2 showed a higher molecular weight than the chitosan recovered using other conventional anti-solvents such as ethanol. The authors concluded that some long chain molecules may be recovered by using compressed CO_2, which made the average molecular weight of the regenerated chitosan more centralized.

The performance of the compressed CO_2 in the precipitation of cellulose from [bmim][OAc] was analyzed and discussed in two different studies.[7,8] Although the yield of the process was lower than the yields of conventional anti-solvents methods (60% *vs.* 90%), the authors showed that yield and degree of polymerization of regenerated cellulose can be easily tuned only by controlling process parameters. Moreover, the regenerated cellulose obtained by compressed CO_2 anti-solvent process presented specific characteristics, which overcomes some drawbacks of the native cellulose or the regenerated cellulose using conventional methods. Morphological characteristics of regenerated cellulose like smoother surface, thicker shape and a more homogeneous texture were reported. Overall, Sun *et al.*, and Liu and co-workers stated that the designed process, combining the tune of yield and DP with energy consumption comparison, is an easy, efficient and more sustainable process for regeneration of cellulose recovered from biomass.

8.3.2.1 Anti-solvent Process Condition Effects

From the reviewed literature, pressure, temperature and reaction time (residence time) and addition of co-solvent are the main factors affecting the biopolymers precipitation from a molecular solvent or IL-solution. The major effects of these parameters in the process yield and regenerated biopolymers characteristics are described in the following sections.

8.3.2.1.1 Pressure, Temperature and Reaction Time. Haimer and co-workers explored the precipitation of chitin and hemicellulose from DMSO[14] and DMSO : water mixtures. The authors showed that pressure variations close to critical pressure promotes changes in mass transfer between solvent and anti-solvent (mass transfer resistance) and revealed to play an important role in adjusting the size and morphology of the precipitating hemicellulose particles. Above critical pressure, mass transfer resistance is reduced to a minimum and small particles of uniform size distribution were obtained. However, agglomerates of larger particles were produced by decreasing the precipitation pressure below the critical value.

Sun and co-workers, studied the precipitation of chitosan and cellulose from ([bmim][OAc]) solutions using compressed CO_2.[8,9] They concluded that the yield of regenerated biopolymers and its degree of polymerization (DP) can be easily tuned by controlling the process parameters, in particular temperature, pressure and reaction time. For both biopolymers, they observed that at 25 °C, the yield of regeneration, reaches a maximum (approximately 60%) at a pressure of 6.6 MPa and for a reaction time of 3 and 5 h for cellulose and chitosan, respectively.

In the case of cellulose, for pressures below 6.6 MPa, the yield of regeneration increased remarkably with increasing CO_2 pressure and decreases with increasing temperature. Similarly, the DP showed a decreasing trend, diminishing gradually for higher pressure and reaction time.

At a certain pressure, higher temperature promotes lower CO_2 density and solvation strength, reducing the cellulose extraction. In this case, with short chain polymers have difficulties in precipitate from the system and the average DP of the regenerated cellulose tends to increase.[8] This phenomenon is one of the advantages of using compressed CO_2 as anti-solvent. Contrarily to the anti-solvent reaction in ethanol, cellulose with different DP can be separated. DP of the precipitated biopolymers, in particular cellulose, can be controlled by the pressure and temperature of the compressed gas. Nevertheless, in comparison with the conventional anti-solvents used to regenerate cellulose (water, ethanol and acetonitrile), the compressed CO_2 anti-solvent process led to a relatively lower yields of regenerated cellulose (yield of about 60% compared with yields above 90%).[7]

8.3.2.1.2 Influence of Co-solvent. The addition of a co-solvent has an important impact on the regeneration process by anti-solvent method once the solubility of the biopolymer on the molecular solvent or IL-solution can be greatly improved.

One example is the work of Haimer *et al.*, where they evaluated the use of DMSO : water mixtures instead of pure DMSO to solubilize hemicellulose. The addition of water allowed the processing of larger amounts of hemicelluloses due to the increased solubility but also led to changes in organic solvent/scCO_2 properties. Mass transfer resistance increases in the presence of water, producing in the end larger hemicellulose aggregates. This effect promotes the use of water to tune the particle size and morphology.

Another more common application of co-solvent is the addition of aprotic polar solvents to ILs-solutions to increase the dissolution of cellulose and delignification of lignocellulosic biomass.[3] It is known that the presence of an aprotic polar solvent will induce the dissociation of IL producing more solvated cations and free anions, available to interact with celluloses and other biopolymers.[9,24] Based on this idea, Sun and co-workers investigated the effect of the addition of aprotic solvents to IL-solutions on the precipitation of cellulose using compressed CO_2 anti-solvent method. Three aprotic polar solvents including, DMSO, DMI (1,3-dimethyl-2-imidazolidinone) and DMF (*N,N*-dimethylformamide) were selected to add to the system ([bmim][OAc]) : co-solvent : microcrystalline cellulose (10% wt). The results demonstrated that for the same reaction time and for a compressed CO_2 at 6.6 MPa and 25 °C, the application of the co-solvent led to a higher recovery of the cellulose compared with the neat system IL : cellulose. Moreover, authors observed that the addition of the co-solvent promoted an increase in the reaction rate. The yield of the regenerated cellulose achieved a maximum after only 2 h of reaction time (compared with the 3 h for the neat system). Another benefit demonstrated for the use of co-solvent is the capacity of fine tuning of the final DP of the regenerated cellulose. For a pressure range between 6.6–20 MPa, the DP values obtained for the regenerated cellulose were in the narrow range of 157–175.[25] This is associated to the fact that the addition of any of the investigated aprotic polar solvent to the system [bmim][OAc] : cellulose prompt the expansion upon the dissolution of compressed CO_2. DMSO was the aprotic polar solvent which presents the higher ratio of volume of expansion to dipole moment.[26]

8.3.2.1.3 Volume of Expansion. Sun *et al.* and Liu and co-workers also explored the influence of changes on volume of expansion on the precipitation of chitosan and cellulose.[7–9] The loading of compressed CO_2 to the solution containing the biopolymer, may cause volume expansion which will promote a longer distance between biopolymer and solvent contributing for their hydrogen bonding disruption. The biopolymer may be surrounded abundantly by CO_2 that favours the separation from the solvent. From this point of view, large biopolymers, with long chain and large molecular weight, have more probabilities to be surrounded by the compressed CO_2 and to be precipitated easily.

8.3.2.1.4 Solvatochromic Effects. The use of the solvatochromic method for studying the effects of solvents in supercritical fluids is nowadays very effective. Taft and his team presented a work tackling this effect[27–31] measuring the solvatochromic shifts of many solute molecules in a wide range of liquid solvents. One of the major contributions was development of a π^* solubility scale which encompasses a wide range of solute types and which includes several solvent properties, including polarity, polarizability, hydrogen bonding ability, *etc.* In the study of Sun *et al.*[8] the precipitation of cellulose from ILs using compressed CO_2 as an anti-solvent to obtain the products with different DP values was investigated. The employed IL was the [bmim][OAc], which is one of the most efficient ILs for the dissolution of biomass.[32] The effects of solvatochromic parameters of these systems by compressed CO_2 were also investigated to explore the possible anti-solvent precipitation mechanism. In this work the absorbance spectra of UV-vis of [bmim][OAc] and three different mixtures of aprotic polar solvents were recorded. The study was conducted under different pressures. To calculate solvatochromic parameters the maximum absorbance length of the solvents was used. In this case, the authors used the solvatochromic Kamlet–Taft parameters because they are one of the simplest parameters and are frequently used for measurements of the solvent properties. Kamlet–Taft parameters are dipolarity/polarizability (π^*), hydrogen-bonding acidity (α) and hydrogen-bonding basicity (β). These can be calculated from the wavelength shifts of 3–6 different solvatochromic dyes in the solvent, usually including Reichard's dye, nitroaniline and diethylnitroaniline.[6,32] Previous studies indicate that the addition of CO_2 to pure [bmim][OAc] produce polarity/polarizability decrease (π^*). However, no significant changes in the basicity (β) and the acidity (α) of the hydrogen were produced. π^* parameter of [bmim][OAc]/co-solvent decreases inversely proportionally to CO_2 pressure ratio. This behaviour is like the [bmim][OAc] in absence of co-solvents. The π^* value of [bmim][OAc]/DMSO mixture undergoes a small decrease when the pressure of CO_2 increases. This increase is significantly higher than those of the mixture with DMF or DMI. In both cases, the value was smaller than the value of [bmim][OAc]. The polarity of the system using aprotic solvents did not suffered important changes. This phenomenon is used to explain the narrow molecular weight distribution and the degree of polymerization values of regenerated cellulose using compressed CO_2.

In this research, the effect of CO_2 in the values of β, α and (β–α) parameters for [bmim][OAc]/co-solvent was also studied. As mentioned above β value represents bond accepting ability, α donating ability and (β–α) represents net basicity. The β parameter value is greatly influenced by the used co-solvent. The effect of the CO_2 pressure on the value of β parameter has no relevance in any of the analyzed systems. Instead, the α parameter values increased directly proportionally to the pressure of CO_2 (compared with [bmim][OAc]). The most significant case was when DMF was used as co-solvent. When combined with β values, a similar order can be obtained as the order of π^*. This suggests that CO_2 molecules can react, to some extent, with IL ions through hydrogen bonds. High values of (β–α) observed by the authors indicate the high capacity to cellulose dissolution. This high capacity results in an incomplete precipitation of cellulose using compressed CO_2 as anti-solvent.

Therefore, varying the load of CO_2 in such systems is possible to reduce the polarity and net donating ability of hydrogen bonds. This indicates that the CO_2 may occupy the initial space between cations and anions of IL instead of cellulose. In addition, regenerated cellulose obtained employing IL and compressed CO_2 as anti-solvent has a certain specificity obtaining better results than conventional anti-solvent systems.

The same parameters were studied in other research for the recovery of chitosan.[9] Considering that the presence of CO_2 can modify the mass transport properties of IL, Kamlet–Taft parameters were calculated. These parameters were studied for the mixture of [bmim][OAc] and CO_2 as functions of pressure based on the solvatochromic behaviour of the indicator *N*,*N*-diethyl-4-nitroaniline (DENA), 4-nitroaniline (NA) and Reichard's dye 33.

In the case of chitosan, a noticeable sharp decline in π^* parameter after adding CO_2 at 6.6 MPa. However, this value was slightly reduced when the pressure decreased from that pressure. This is due to the changes in the polarity of the system and the dipole–dipole and dipole-induced dipole interactions generated between cations and anions in the system. This corroborates that a small molecule such as CO_2 is able to occupy the volume between the cations and anions producing interactions with the dye molecules. The study for β and α parameters at different CO_2 pressures did not show significant changes. Instead, a decreasing trend in the (β–α) parameter of [bmim][OAc] is in agreement with other experimental results obtained for different ILs in other ref. 33.

8.3.3 Mechanisms of Precipitation

Barber and co-authors demonstrated that the chemisorption of CO_2 is a possible mechanism for coagulation of chitin and cellulose dissolved in 1-ethyl-3-methylimidazolium acetate [emim][OAc] using supercritical and gashouse CO_2 through the formation of the zwitterionic imidazolium carboxylate. The authors concluded that carboxylate zwitterions formed upon addition of compressed CO_2, may be responsible for sequestering acetate anions from the system leading to the precipitation of the biopolymer, chitin or cellulose (Scheme 8.4).

The same mechanism may occur during the precipitation of chitosan or cellulose from [bmim][OAc] upon the addition of compressed CO_2, described by Sun and co-workers in 2014.[8,9] They demonstrated that at 6.6 MPa and 25 °C, compressed CO_2 can be significantly dissolved into the chitosan/[bmim][OAc] solution and strongly interact with the cations and anions of the IL. Later, the same author, based on previous works, theorized that the

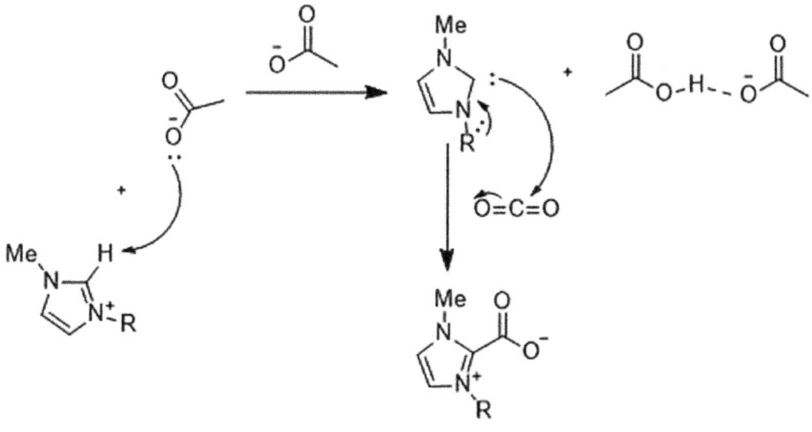

Scheme 8.4 The mechanism of formation of carboxylate zwitterions imidazolium-2-carboxylate ([bmim][COO]).
Reprinted from Z. Liu, X. Sun, M. Hao, C. Huang, Z. Xue, T. Liu, Preparation and characterization of regenerated cellulose from ionic liquid using different methods,[7] *Carbohydr. Polym.*, **117**, 99–105, Copyright (2015), with permission from Elsevier.

precipitation of the biopolymers such as cellulose or chitosan from the ILs solution may be due to the disruption of the hydrogen bonding between biopolymer and IL caused by the formation of the carboxylate zwitterions.[7] To explore this hypothesis, Liu and co-workers compared NMR spectra of pure [bmim][OAc] and [bmim][OAc] upon contact with compressed CO_2 at 6.6 MPa for 3 h. They found new ^1H resonances at 3.90 ppm and 4.42 ppm, which are endorsed to the formation of carboxylate zwitterions imidazolium-2-carboxylate ([bmim][COO]). Additionally, the authors also observed a new ^{13}C resonance (152.85 ppm) upon addition of compressed CO_2, which they clearly correlate with the presence of the compressed gas. Once the isolated CO_2 presents a single resonance at 125.4 ppm, the new identified resonance cannot be attributed to CO_2 molecules dissolved into acetate-based ILs[7] but to a new specie formed from the reactions between CO_2 and the ILs as previously reported by Barber *et al.* as the chemisorption phenomena.

8.4 Perspectives

It is a fact that biopolymers from biomass (lignocellulosics and marine) may provide an important alternative to polymers derived from petroleum. Biopolymers are mainly biocompatible and often exceed the properties of the polymers obtained from fossil sources. Currently, biomass sources are generally cheaper than fossil fuels but the processes of obtaining polymers from fossil sources are so optimized that they are still great competitors. The development of new methodologies or improved processes for efficient and economic utilization as well as the conversion of these biopolymers is still challenging as it is a key requirement for the gradual replacement of petroleum-derived polymers.

Numerous researchers have suggested different ways to recover biopolymers from biomass. The recovery yields obtained using these technologies are quite high, obtained biopolymers have a high quality and the technology is highly selective. But techniques employing the combination of DMSO and ILs have a major disadvantage in terms of recovery and separation of solvents. Furthermore the use of organic solvents such as DMSO increases the cost of the process and therefore the price of the final product. Moreover, the technology in which the anti-solvent effect using $scCO_2$ is used is considered fully ecological and environmentally friendly as it avoids the use of organic compounds. The products used for the recovery of biopolymers from biomass are fully recoverable and reusable; because of this anti-solvent compressed CO_2 technology can be considered a highly efficient methodology. A stronger investment on this technology would reduce the price of the recovery processes of biopolymers. This price decrease would also be reflected in the value of final products, which would make those ecologically obtained biopolymers great competitors for the materials obtained from fossil sources. The use of pressurized CO_2 as anti-solvent for the regeneration of biomass has not been studied extensively and in detail yet. Therefore, it is a field with great potential for study varying thermodynamic factors of reaction, chemical composition of the solvents and other physical conditions. Combining the ever-increasing technological development, scientific progress and public investment in techniques for obtaining sustainable materials, a great process of obtaining biocompatible and biodegradable materials with several applications could be achieved.

References

1. A. Montes, M. D. Gordillo, C. Pereyra and E. J. Martínez de la Ossa, *Chem. Eng. Technol.*, 2014, **37**, 141.
2. A. Montes, N. Kin, M. D. Gordillo, C. Pereyra and E. J. M. de la Ossa, *J. Supercrit. Fluids*, 2014, **89**, 58.
3. A. R. Xu, J. J. Wang and H. Y. Wang, *Green Chem.*, 2010, **12**, 268–275.
4. L. Pérez-Álvarez, J. Laza and A. Álvarez-Bautista, *Curr. Pharm. Des.*, 2016, **22**, 3380.
5. M. Petkovic, K. R. Seddon, L. P. N. Rebelo and C. S. Pereira, *Chem. Soc. Rev.*, 2011, **40**, 1383–1403.
6. A. W. King, J. Asikkala, I. Mutikainen, P. Järvi and I. Kilpeläinen, *Angew. Chem.*, 2011, **123**, 6425–6429.
7. Z. Liu, X. Sun, M. Hao, C. Huang, Z. Xue and T. Mu, *Carbohydr. Polym.*, 2015, **117**, 99.
8. X. Sun, Y. Chi and T. Mu, *Green Chem.*, 2014, **16**, 2736–2744.
9. X. Sun, Z. Xue and T. Mu, *Green Chem.*, 2014, **16**, 2102.
10. A. M. da Costa Lopes, K. G. João, A. R. C. Morais, E. Bogel-Lukasik and R. Bogel-Lukasik, *Sustainable Chem. Processes*, 2013, **1**, 3.
11. G. J. Kabo, A. V. Blokhin, Y. U. Paulechka, A. G. Kabo, M. P. Shymanovich and J. W. Magee, *J. Chem. Eng. Data*, 2004, **49**, 453–461.
12. Y. U. Paulechka, G. J. Kabo, A. V. Blokhin, O. A. Vydrov, J. W. Magee and M. Frenkel, *J. Chem. Eng. Data*, 2003, **48**, 457–462.
13. U. Domanska and R. Bogel-Lukasik, *J. Phys. Chem. B*, 2005, **109**, 12124–12132.
14. E. Haimer, M. Wendland, A. Potthast, U. Henniges, T. Rosenau and F. Liebner, *J. Supercrit. Fluids*, 2010, **53**, 121.
15. D. A. Fort, R. C. Remsing, R. P. Swatloski, P. Moyna, G. Moyna and R. D. Rogers, *Green Chem.*, 2007, **9**, 63–69.
16. A. V. M. Nunes and C. M. M. Duarte, *Materials*, 2011, **4**, 2017–2041.
17. K. Wu and J. Li, *J. Supercrit. Fluids*, 2008, **46**, 211.
18. N. Foster, R. Mammucari, F. Dehghani, A. Barrett, K. Bezanehtak, E. Coen, G. Combes, L. Meure, A. Ng, H. L. Regtop and A. Tandya, *Ind. Eng. Chem. Res.*, 2003, **42**, 6476.
19. A. Martín and M. J. Cocero, *Adv. Drug Delivery Rev.*, 2008, **60**, 339.
20. E. J. Beckman, *J. Supercrit. Fluids*, 2004, **28**, 121–191.
21. M. Mukhopadhyay and S. V. Dalvi, *J. Supercrit. Fluids*, 2004, **29**, 221.
22. A. E. Andreatta, L. J. Florusse, S. B. Bottini and C. J. Peters, *J. Supercrit. Fluids*, 2007, **42**, 60.

23. P. S. Barber, C. S. Griggs, G. Gurau, Z. Liu, S. Li, Z. Li, X. Lu, S. Zhang and R. D. Rogers, *Angew. Chem.*, 2013, **125**, 12576–12579.
24. Y. Zhao, X. Liu, J. Wang and S. Zhang, *J. Phys. Chem. B*, 2013, **117**, 9042.
25. R. Yan, F. L. Ju, H. Y. Wang, C. H. Sun, H. Q. Zhang, M. Y. Shao and Y. Z. Wang, *Anal. Methods*, 2014, **6**, 9561–9566.
26. R. Fernández-Prini, in *Physical Chemistry of Organic Solvent Systems*, eds. A. K. Covington and T. Dickinson, Plenum Press, London, U.K., 1973.
27. M. J. Kamlet and R. W. Taft, *J. Am. Chem. Soc.*, 1976, **98**, 377–383.
28. R. W. Taft and M. J. Kamlet, *J. Am. Chem. Soc.*, 1976, **98**, 2886–2894.
29. T. Yokoyama, R. W. Taft and M. J. Kamlet, *J. Am. Chem. Soc.*, 1976, **98**, 3233–3237.
30. M. J. Kamlet, J. L. Abboud and R. W. Taft, *J. Am. Chem. Soc.*, 1977, **99**, 6027.
31. M. E. Jones, R. W. Taft and M. J. Kamlet, *J. Am. Chem. Soc.*, 1977, **99**, 8452.
32. R. Pezoa, V. Cortinez, S. Hyvarinen, M. Reunanen, J. Hemming, M. E. Lienqueo, O. Salazar, R. Carmona, A. Garcia, D. Y. Murzin and J. P. Mikkola, *Cellul. Chem. Technol.*, 2010, **44**, 165–172.
33. J. Lu, C. L. Liotta and C. A. Eckert, *J. Phys. Chem. A*, 2003, **107**, 3995.

CHAPTER 9

Perspectives of the Development of High-pressure Technologies in Biomass Processing

RAFAL M. LUKASIK

Unidade de Bioenergia, Laboratório Nacional de Energia e Geologia, I.P., Estrada do Paço do Lumiar 22, 1649-038 Lisboa, Portugal
Email: rafal.lukasik@lneg.pt

9.1 Perspectives

Today's world economy requires increasing energy resources and fossil fuel sources are not able to fulfil this demand in a sustainable manner. Furthermore, geopolitical instability and concerns regarding the environmental impacts of current technologies have resulted in a global shift towards new and alternative technologies offering more sustainable sources of energy, materials, chemicals and value-added commodities. One of the most sustainable resources is biomass, especially lignocellulosic biomass. However, the delivery of a vast range of commodities employing sustainable supply chains is still challenging for a bioeconomy. The track to achieve it is a biorefinery concept. The biorefinery concept considers the use of biomass (*e.g.* lignocellulosic) as a feedstock for diverse industries (Figure 9.1).

Following a definition adopted by the International Energy Agency Bioenergy Task 42 Biorefining is the sustainable processing of biomass into a spectrum of marketable products and energy.[1] In other words, the biorefinery is a term used to define industrial facilities that cover an extensive

Green Chemistry Series No. 48
High Pressure Technologies in Biomass Conversion
Edited by Rafał M. Łukasik

Published by the Royal Society of Chemistry, www.rsc.org

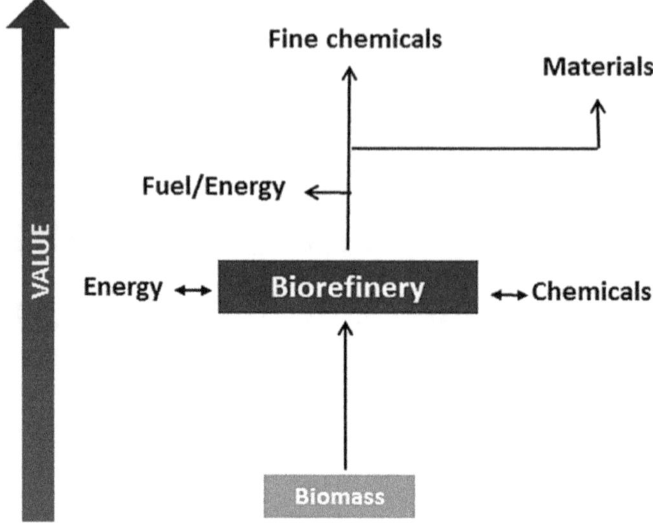

Figure 9.1 The biorefinery concept.

range of combined technologies in which biomass is transformed and converted, in a sustainable manner, into a wide range of value-added products leading to direct similarities to today's petrorefineries.[2] Hence, according to this concept, biomass would be a source of diverse commodities produced by comprehensive processing of biomass by cascade valorization of each fraction.

Considering the worldwide production of lignocellulosic biomass estimated on 10^{10} million tons,[7] biomass reveals to be a real opportunity for novel fossil-alternative technologies. Yet, intricate and recalcitrant as well as diverse matrix of biomass makes the conversion of this feedstock into diverse commodities very challenging and complex. The manufacturing of biofuels or other value-added products from these kinds of biomasses encompasses several steps including hydrolysis of polysaccharides and their further conversion, separation of lignin from the residue and purification of the final products. Despite dozens of years of research about the biomass processing the valorization technologies are still mostly characterized by low yields and high costs, which burden the production of commodities at competitive costs. Thus, the choice of the technology employed to be used, as a biomass processing method, should be always based on the characteristics and properties of available feedstock and on the impact of the process on the further valorization pathway by, for instance, the production of toxic compounds for the metabolism by microorganisms, energy demand and waste production.[2] Hence, it can be concluded that there is no ideal method of biomass processing nevertheless the biomass should be processed in the greenest possible way to avoid a possibility to lose the environmental and social benefits related to the processed feedstock.

Therefore, the use of greener and more sustainable methods of biomass processing *e.g.* with high-pressure fluids are one of the future and promising methods to be employed in the delivery of more sustainable commodities. Among high-pressure fluids, water is considered as potentially interesting fluid for biomass treatment.[3–9] Second, the most interesting fluid is CO_2[2,10–13] and when mixed with water can generate an additional acidic environment favouring biomass hydrolysis.[2,14] According to Sheldon "*(...) the use of water and supercritical CO_2 as reaction media is also consistent with the current trend towards the use of renewable, biomass-based raw materials (...)*".[15] This confirms that either water or CO_2 will securely play an important role in novel methodologies of biomass processing.

The results presented in this book shows that there is still a room for improvement in biomass processing with high-pressure fluids and the use of CO_2 should be explored further in several areas of lignocellulosic feedstock conversion. One of these areas is a direct hydrolysis of untreated biomass in the presence of CO_2 and possible integration and intensification of the biomass valorization. In an ideal situation, biomass feedstock would not require a cost- and energy-demanding pre-treatment step. Additionally, the produced liquid streams would contain easily fermentable sugars to obtain biofuels and/or value added products. The current research shows that it is possible at least for pure cellulose samples.[16,17] Hence, there is still a possibility to extend this knowledge to more complex matrix of lignocellulosic biomass.

Other challenge related to engagement of high-pressure fluids in lignocellulosic biomass is the limited solubility of organic compounds in water. Taking into account also limited mutual solubility of CO_2 and water the advantages of sub- or supercritical conditions requires high-pressure and temperatures. Therefore to benefit from the CO_2 properties as interesting fluid, solvents alternative to water could be explored. One of such examples can be carbonate-type solvents, *e.g.* 1,2-glycerolcarbonate.[18,19] A possible manufacturing of such a solvent by direct reaction between excessing glycerol and CO_2, makes this solvent an interesting alternative for biomass processing.

One of the very little studied fields in biomass processing, in general, is the valorization of lignin. Nowadays the lignin in the industry serves for energy supply to the reactive system and at the R&D level lignin is considered as source of polyphenols.[20–24] Even so the research in this field is still very scarce. The reason for this is a very complex nature of lignin which depends on the biomass processed but also can be altered by the process (storage, pre-treatment) conditions. This causes a lack of typical patterns to process this biomass fraction turning it very difficult. Another "problematic" fraction of biomass is hemicellulose. The heterogeneity (lower than in case of lignin) and complexity of hemicellulose is dependent on the biomass type and other external factors (*e.g.* climacteric and storage conditions). However, this diversity can be considered an opportunity because of numerous products that can be obtained, depending on the processing technique applied.[11,25–28]

Passing from the chemical reaction level to a technology system level, one of the crucial aspect in the further development of new processing methods (including those with high-pressure fluids) is the need of a fair comparison between the current existing technologies of biomass processing and the novel methods presented in this book. One of the empirical methods of such a comparison would be a SWOT (Strengths, Weaknesses, Opportunities and Threats) analysis, which helps to depict the pros and cons of each technology. Of course, more complex Life Cycle Analysis approaches can and should be scrutinized as well as they would provide a factual comparison between industrial biomass processing realities and novel methods that are often still in their infancy. The CO_2–water technologies are still at low technical readiness level (TRL)[29] of 4–5 (Figure 9.2), *i.e.* from technology validated in laboratory (TRL 4) to technology validated in relevant environment (industrially relevant environment in the case of key enabling technologies) (TRL 5). Hence, the comparisons to industrial scenarios are needed to be done in the way that considers such differences in TRLs between current and novel methods.

Often claimed as a burden for the employment of high-pressure technologies, is the need of specialized equipment resistant to high-pressures and temperatures. Definitively, it is one of the drawbacks of high-pressure processing of biomass, however, it must be stated that ultra-high pressure processing[30] (HPP) of food has been an industrial reality for many decades. Additionally, it should be stated again that the potential gains from the use of lower temperatures rather than classical technologies and possibility to avoid corrosive chemicals (*e.g.* mineral acids) is definitively beneficial for

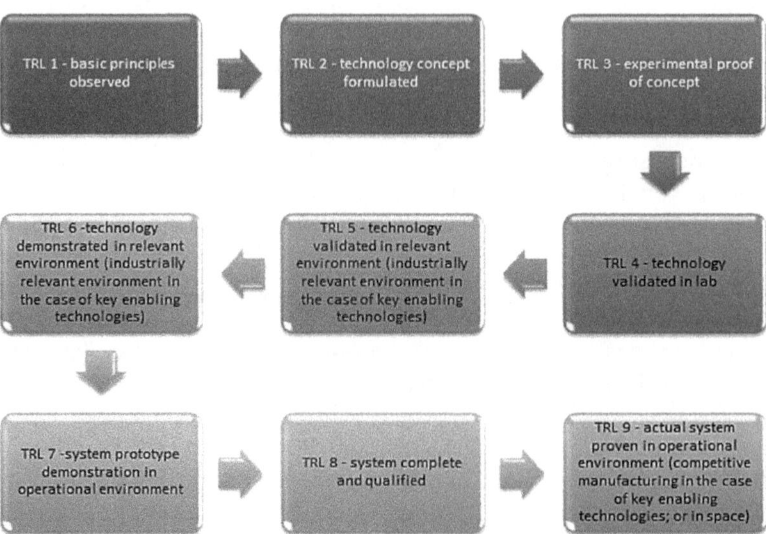

Figure 9.2 The Technical Readiness Level (TRL) scale.

such technologies. Therefore it can be concluded that although high-pressure processes would probably generate high capital investment costs (CAPEX) but operational costs (OPEX) would be definitively more favourable for technologies involving CO_2–water than energy-demanding obsolete technologies such as steam explosion.[31,32] Thus, again, only a comprehensive analysis of the pros and cons with adequate sustainability analysis (economic, environmental and social aspects) of current and novel technologies allows to provide the answer about the feasibility of those technologies for particular industrial scenarios.

This book shows that high-pressure CO_2 and CO_2–H_2O systems have proven beneficial for the conversion of biomass in several processes.[2,10,11,13,14,21,26,33–69] The ability of high-pressure CO_2 and CO_2–H_2O systems to engineer reaction conditions, reduce energy requirements and tune reaction rates, product selectivity and catalyst activity by manipulating only pressure, temperature or CO_2 content is especially attractive. For these reasons, there is continued interest in developing these CO_2-based processes in the context of biorefineries. The biggest obstacle to overcome is to demonstrate that the development of these technologies will be environmentally and economically worthwhile. These high-pressure processes have to overcome high capital costs, concerns over safety and the energetic cost of CO_2 conditioning within the process. Therefore, future research directions should focus on the development of highly integrated processes coupled with careful economic and environmental modelling. Such approaches could lead to higher projected process efficiency and thus help to convince the private sector to implement these technologies at a scale that will justify the capital investment and the cost of specialized equipment and engineers. Small improvements in efficiencies and the accuracy of predictions for integrated process models will make CO_2-based processes more cost-effective and attractive to industry. Such improvements will likely depend on a better understanding of the behaviour of CO_2–water mixtures in the presence of the many biomass derived molecules that will be solvated by these systems. The effect of these molecules, especially at high concentrations is still very poorly understood. In parallel, political incentives such as carbon tax credits associated with the use of green and renewable solvents and catalysts could further encourage the adoption CO_2-based processes. On the other hand replacement of the traditional organic solvents with "green" solvents has a direct impact on the chemical side of the synthesis process, which require a deeper understanding on the reaction system as well as the process mechanism.

9.2 Conclusions

An awareness of changes occurring in today's world is driving the necessity to substitute the current hazardous technologies by more sustainable, greener and environmentally friendly processes. Lignocellulosic biomass, due to its high abundance, seems to be a potential candidate for a sustainable carbon source. High-pressure fluids including carbon dioxide and water as well as other nowadays broadly examined solvents fit into a trend of novel processes with a great potential for the future. High-pressure CO_2 has been demonstrated to be very effective in fractionation and hydrolysis of lignocellulosic fractions, hydrolysis, extraction and downstream processing, showing also it utility to achieve a range of goals in biomass processing. Therefore there are numerous arguments for using CO_2 and other high-pressure fluids within biorefineries but the ultimate implementation of these technologies will depend on multiple research efforts involving experimental, computational and policy efforts.

Acknowledgements

This chapter was written with substantial help of authors of other chapters and special thanks are addressed to Jeremy S. Luterbacher, Juan García-Serna and Maya Chatterjee. This work was supported by the Fundação para a Ciência e Tecnologia (FCT, Portugal) through grant IF/00424/2013.

References

1. R. Van Ree, *IEA Bioenergy Task 42 Biorefining*, International Energy Agency - IEA Bioenergy, Wageningen UR – Food and Bio-based Research, 2014.
2. A. R. C. Morais, A. M. da Costa Lopes and R. Bogel-Lukasik, *Chem. Rev.*, 2015, **115**, 3–27.
3. E. Ruiz, C. Cara, P. Manzanares, M. Ballesteros and E. Castro, *Enzyme Microb. Technol.*, 2008, **42**, 160–166.
4. A. Romani, G. Garrote, J. L. Alonso and J. C. Parajo, *Bioresour. Technol.*, 2010, **101**, 8706–8712.
5. C. Cara, E. Ruiz, M. Ballesteros, P. Manzanares, M. J. Negro and E. Castro, *Fuel*, 2008, **87**, 692–700.
6. H. D. Zhang, S. H. Xu and S. B. Wu, *Bioresour. Technol.*, 2013, **143**, 391–396.
7. Q. Yu, X. S. Zhuang, Z. H. Yuan, W. Qi, W. Wang, Q. Wang and X. S. Tan, *Bioresour. Technol.*, 2013, **144**, 210–215.
8. N. Mosier, R. Hendrickson, N. Ho, M. Sedlak and M. R. Ladisch, *Bioresour. Technol.*, 2005, **96**, 1986–1993.
9. M. Moller, P. Nilges, F. Harnisch and U. Schroder, *ChemSuschem*, 2011, **4**, 566–579.
10. G. P. van Walsum, *Appl. Biochem. Biotechnol.*, 2001, **91–93**, 317–329.
11. A. R. C. Morais, A. C. Mata and R. Bogel-Lukasik, *Green Chem.*, 2014, **16**, 4312–4322.
12. R. Marriott and E. Sin, in *The Role of Green Chemistry in Biomass Processing and* Conversion, ed. H. Xie and N. Gathergood, John Wiley & Sons, Inc., 2013, pp. 181–204.
13. A. Demirbas, *Energy Convers. Manage.*, 2001, **42**, 279–294.

14. G. P. van Walsum and H. Shi, *Bioresour. Technol.*, 2004, **93**, 217–226.
15. R. A. Sheldon, *Green Chem.*, 2014, **16**, 950–963.
16. G. Muratov, K. W. Seo and C. Kim, *J. Ind. Eng. Chem.*, 2005, **11**, 42–46.
17. Y. Z. Zheng and G. T. Tsao, *Biotechnol. Lett.*, 1996, **18**, 451–454.
18. G. V. S. M. Carrera, Z. Visak, R. Bogel-Lukasik and M. N. da Ponte, *Fluid Phase Equilib.*, 2011, **303**, 180–183.
19. J. George, Y. Patel, S. M. Pillai and P. Munshi, *J. Mol. Catal. A: Chem.*, 2009, **304**, 1–7.
20. J. E. G. van Dam, B. de Klerk-Engels, P. C. Struik and R. Rabbinge, *Ind. Crops. Prod.*, 2005, **21**, 129–144.
21. R. J. A. Gosselink, W. Teunissen, J. E. G. van Dam, E. de Jong, G. Gellerstedt, E. L. Scott and J. P. M. Sanders, *Bioresour. Technol.*, 2012, **106**, 173–177.
22. Z. S. Yuan, S. N. Cheng, M. Leitch and C. B. Xu, *Bioresour. Technol.*, 2010, **101**, 9308–9313.
23. T. L. K. Yong and Y. Matsumura, *Ind. Eng. Chem. Res.*, 2012, **51**, 11975–11988.
24. A. J. Ragauskas, C. K. Williams, B. H. Davison, G. Britovsek, J. Cairney, C. A. Eckert, W. J. Frederick, J. P. Hallett, D. J. Leak, C. L. Liotta, J. R. Mielenz, R. Murphy, R. Templer and T. Tschaplinski, *Science*, 2006, **311**, 484–489.
25. F. M. Girio, C. Fonseca, F. Carvalheiro, L. C. Duarte, S. Marques and R. Bogel-Lukasik, *Bioresour. Technol.*, 2010, **101**, 4775–4800.
26. B. C. Saha, *J. Ind. Microbiol. Biotechnol.*, 2003, **30**, 279–291.
27. E. Guerra-Rodriguez, O. M. Portilla-Rivera, L. Jarquin-Enriquez, J. A. Ramirez and M. Vazquez, *Biomass Bioenergy*, 2012, **36**, 346–355.
28. A. Garde, G. Jonsson, A. S. Schmidt and B. K. Ahring, *Bioresour. Technol.*, 2002, **81**, 217–223.
29. *Technology readiness level (TRL), HORIZON 2020-Work Programme 2014–2015 General Anneses, Extract from Part 19-Commission Decision C(2014)4995*, European Commission, Brussels, Belgium, 2014.
30. M. J. Mota, R. P. Lopes, I. Delgadillo and J. A. Saraiva, *Biotechnol. Adv.*, 2013, **31**, 1426–1434.
31. I. Ballesteros, M. Ballesteros, C. Cara, F. Saez, E. Castro, P. Manzanares, M. J. Negro and J. M. Oliva, *Bioresour. Technol.*, 2011, **102**, 6611–6616.
32. J. A. Pérez, I. Ballesteros, M. Ballesteros, F. Sáez, M. J. Negro and P. Manzanares, *Fuel*, 2008, **87**, 3640–3647.
33. H. D. Zhang and S. B. Wu, *Bioresour. Technol.*, 2014, **158**, 161–165.
34. C. Zetzl, K. Gairola, C. Kirsch, L. Perez-Cantu and I. Smirnova, *Chem. Ing. Tech.*, 2011, **83**, 1016–1025.
35. J. Z. Yin, L. D. Hao, W. Yu, E. J. Wang, M. J. Zhao, Q. Q. Xu and Y. F. Liu, *Chin. J. Catal.*, 2014, **35**, 763–769.
36. O. Yemis and G. Mazza, *Bioresour. Technol.*, 2011, **102**, 7371–7378.
37. S. X. Wu, H. L. Fan, Y. Xie, Y. Cheng, Q. A. Wang, Z. F. Zhang and B. X. Han, *Green Chem.*, 2010, **12**, 1215–1219.

38. G. P. Van Walsum, M. Garcia-Gil, S.-F. Chen and K. Chambliss, *Appl. Biochem. Biotechnol.*, 2007, 301–311.
39. S. K. Thangavelu, A. S. Ahmed and F. N. Ani, *Appl. Energy*, 2014, **128**, 277–283.
40. N. Srinivasan and L. K. Ju, *Biomass Bioenergy*, 2012, **47**, 451–458.
41. M. H. L. Silveira, B. A. Vanelli, M. L. Corazza and L. P. Ramos, *Bioresour. Technol.*, 2015, **192**, 389–396.
42. R. A. Sheldon, *Catal. Today*, 2011, **167**, 3–13.
43. C. Schacht, C. Zetzl and G. Brunner, *J. Supercrit. Fluids*, 2008, **46**, 299–321.
44. P. E. Savage, S. Gopalan, T. I. Mizan, C. J. Martino and E. E. Brock, *AIChE J.*, 1995, **41**, 1723–1778.
45. T. Sato, T. Furusawa, Y. Ishiyama, H. Sugito, Y. Miura, M. Sato, N. Suzuki and N. Itoh, *Ind. Eng. Chem. Res.*, 2006, **45**, 615–622.
46. M. D. A. Saldana and C. S. Valdivieso-Ramirez, *J. Supercrit. Fluids*, 2015, **96**, 228–244.
47. R. M. N. Roque, M. N. Baig, G. A. Leeke, S. Bowra and R. C. D. Santos, *Resour., Conserv. Recycl.*, 2012, **59**, 43–46.
48. T. Rogalinski, K. Liu, T. Albrecht and G. Brunner, *J. Supercrit. Fluids*, 2008, **46**, 335–341.
49. O. Pourali, F. S. Asghari and H. Yoshida, *Chem. Eng. J.*, 2010, **160**, 259–266.
50. M. Osada, N. Hiyoshi, O. Sato, K. Arai and M. Shirai, *Energy Fuel*, 2007, **21**, 1854–1858.
51. P. S. Nigam and A. Singh, *Prog. Energy Combust. Sci.*, 2011, **37**, 52–68.
52. N. Narayanaswamy, A. Faik, D. J. Goetz and T. Y. Gu, *Bioresour. Technol.*, 2011, **102**, 6995–7000.
53. Y. Matsushita, K. Yamauchi, K. Takabe, T. Awano, A. Yoshinaga, M. Kato, T. Kobayashi, T. Asada, A. Furujyo and K. Fukushima, *Bioresour. Technol.*, 2010, **101**, 4936–4939.
54. T. D. Matson, K. Barta, A. V. Iretskii and P. C. Ford, *J. Am. Chem. Soc.*, 2011, **133**, 14090–14097.
55. H. Machida, M. Takesue and R. L. Smith, *J. Supercrit. Fluids*, 2011, **60**, 2–15.
56. H. S. Lv, L. Yan, M. H. Zhang, Z. F. Geng, M. M. Ren and Y. P. Sun, *Chem. Eng. Technol.*, 2013, **36**, 1899–1906.
57. Y. Lu, Q. F. Sun, D. J. Yang, X. L. She, X. D. Yao, G. S. Zhu, Y. X. Liu, H. J. Zhao and J. Li, *J. Mater. Chem.*, 2012, **22**, 13548–13557.
58. H. S. Lu, M. M. Ren, M. H. Zhang and Y. Chen, *Chin. J. Chem. Eng.*, 2013, **21**, 551–557.
59. Y. F. Liu, P. Luo, Q. Q. Xu, E. J. Wang and J. Z. Yin, *Cellul. Chem. Technol.*, 2014, **48**, 89–95.
60. J. Li, Y. Lu, D. J. Yang, Q. F. Sun, Y. X. Liu and H. J. Zhao, *Biomacromolecules*, 2011, **12**, 1860–1867.
61. J. W. King, K. Srinivas, O. Guevara, Y. W. Lu, D. F. Zhang and Y. J. Wang, *J. Supercrit. Fluids*, 2012, **66**, 221–231.

62. J. W. King and K. Srinivas, *J. Supercrit. Fluids*, 2009, **47**, 598–610.
63. S. Karagoz, T. Bhaskar, A. Muto and Y. Sakata, *Fuel*, 2005, **84**, 875–884.
64. M. C. Gutierrez, J. A. Siles, A. F. Chica and M. A. Martin, *Biomass Bioenergy*, 2014, **62**, 93–99.
65. T. Y. Gu, M. A. Held and A. Faik, *Environ. Technol.*, 2013, **34**, 1735–1749.
66. M. A. Gao, F. Xu, S. R. Li, X. C. Ji, S. F. Chen and D. Q. Zhang, *Biosyst. Eng.*, 2010, **106**, 470–475.
67. K. Gairola and I. Smirnova, *Bioresour. Technol.*, 2012, **123**, 592–598.
68. A. F. M. Claudio, A. M. Ferreira, M. G. Freire and J. A. P. Coutinho, *Green Chem.*, 2013, **15**, 2002–2010.
69. M. Chatterjee, T. Ishizaka and H. Kawanami, *Green Chem.*, 2014, **16**, 1543–1551.

Subject Index

acetic acid 39–40, 103–4, 126–7
acid catalysts 39
acid hydrolysis 23, 24, 38–40
activated carbon (C) 141, 150–2
aldol reaction 119–23, 146–7
algae *see* microalgae
alkaline hydrogen peroxide 56
alkanes, from 5-HMF 145–9
alkyltrimethylammonium 141
alumina (Al_2O_3) 141, 146–7
amino acids 88, 101–2
α-amylase 70
anaerobic digestion 18
Andrews, Thomas 1
anti-solvent process 166, 169–78
  co-solvent influence on 174–5
  precipitation 177–8
  pressure, temperature and residence time 173–4
  schematic of 170
  solvatochromic effects 175–7
  volume of expansion 175
aqueous phase reforming 145
Arrhenius equation 118
Arrhenius plot 27
aspen wood 94, 103, 125–7
autohydrolysis 38

batch reactors 67–8, 72, 76
  conditions for using 132–3
  set-ups 142, 144
bio-char 16, 18, 26
bio-oil 16, 17–18, 74
biodiesel 13, 16
  from microalgae 77–8
  purification 21, 25
  synthesis 75–6
bioethanol 13, 14
biofuels 11–14
  from 5-HMF 145–9
  from 5-HMF to 2,5-DMF 149–55
  furfural to 2-MF 155–6
biogas 11
biomass
  definition 11, 137
  first generation 11–13, 16, 84
  hydrolysis 88–90
    biomass-derived polymers 91–110
  second and third generation 11–14, 16, 84
  top value-added compounds from 117
  see also lignocellulosic biomass
biomass conversion 16
  challenges of 10, 182
  CO_2-assisted processes 22–3
  green technology 183
  high-pressure technology vii–viii, 181–6
  pre-treatment viii, 37–61
  in $scCO_2$ 73–8
  see also conversion
biopolymers 165–9, 178–9
  anti-solvent effect 169–78
  see also cellulose; hemicellulose
biorefinery concept 10, 14–18, 115–17, 138, 181

based on $scCO_2$ 60
high-pressure CO_2 and CO_2–H_2O systems 19–30
main processes 15–18
products 15, 16
schematic representation 14
2,5-bis(hydroxylmethyl)furan 153–4
[bmim][OAc] 172–8
BSA protein 101
1-butanol 57, 155

capital investment costs 185
Carbon Capture and Sequestration 4
carbon dioxide
anti-solvent effect of 171–7
dielectric constant 47–9
pH, function of temperature and pressure 96
phase behaviour 41–7, 151–2
phase diagram 140
phase equilibrium 49–51
physicochemical properties 19, 41–7
pressure
and conversion 147–8
conversion and selectivity 158–9
and product distribution 151–2
pressure–composition isotherms 50–1
pressure–density diagrams 43
pressure–solubility parameter diagram 45–6
pressure–sound speed diagram 44
pressure–temperature and pressure–density projections 42
pressure–viscosity diagram 44
pressurizing 59
recycling 133
solubility, function of temperature and pressure 89–90, 124–5
supply 59
see also high-pressure CO_2 and CO_2–H_2O systems; supercritical CO_2
carbon dioxide explosion 21, 23
carbon dioxide pre-treatment 61
applications of 51–6
co-solvents in 57–8
scale-up of 58–60
carbonated subcritical water 124–30
carbonic acid
dissocation 88–9, 90
enzymatic reactions and 71–2
formation 19, 26, 51–3
catalysis 16, 18, 26–30
acid catalysts 39
catalyst characterization 142, 143, 144
in *in situ* synthesis 142, 144–5
see also enzymatic reactions; *individual metal catalysts*
catalytic transfer hydrogenation 150
cationic surfactants 141
cellulose 12, 13, 16, 166–7
α, β and γ 167
anti-solvent conditions 172–7
chemical composition 85–6, 137–8, 167
CO_2-assisted processes using 23, 26–7
ethanol production 53
hydrolysis 26–7, 91–3, 117–23
$scCO_2$ treatment 74–5
cetyltrimethylammonium bromide (CTAB) 141–2
chemical processes
in biorefineries 17–18
in high-pressure CO_2 and CO_2–H_2O systems 21–30
thermochemical 17–18, 21, 26
chitin 168, 172–3
chitosan 168–9, 172–4, 176–8
co-solvents 4
anti-solvent influence of 174–5
in CO_2 pre-treatment 57–8
dielectric constant and 48–9

co-solvents (*continued*)
    phase equilibrium and 49–51
    solubility parameters and 46–7
cobalt catalysts 17
coffee decaffeination 2, 3
combined severity parameter 20, 53, 90, 125
conjugated linoleic acid 68
continuous reactors 72–6
    conditions for using 132–3
conversion
    5-HMF 153–5
    furfural 156
    pressure and selectivity in CO_2 158–9
    and pressure in CO_2 147–8
    problems with lignocellulosics 84
    THFA 157–8
    see also biomass conversion
cork 2, 3
corn cob/stover 75
    chemical composition 85
    delignification 103–5
    hydrolysis 95–6, 125–7
corn starch 99–100
corrosion 5–6
costs 30, 31, 182
    capital investment and operational 185
critical density 41–3
critical isotherm 43
critical locus 49–50
critical point 1–2
    binary system 49
    CO_2 41–3
critical pressure 5
    CO_2 41–2
critical temperature 1, 5
    CO_2 41–3
crops
    edible 11–13
    non-edible 11, 12, 13–14
    see also individual crops
degree of polymerization 166, 168, 174
dehydration 18, 24, 119–23, 146–8
dehydrodeoxygenation 26, 28
delignification 57, 102–6
depolymerization 24, 29–30, 101, 106–9
dielectric constant 5, 19, 47–9, 130
diffusion coefficient 43
1,3-dimethyl-2-imidazolidinone 174, 176
dimethyl ether 78
N,N-dimethylformamide 174, 176
2,5-dimethylfuran (2,5-DMF) 28, 29
    from 5-HMF 149–55
dimethylsulfoxide (DMSO) 169, 171–6
2,5-dimethyltetrahydrofuran (DMTHF) 151–2, 153–4
diols 139, 156
1,4-dioxane 102–4
dissociation constant 118–19
dissolution step 169
2,5-DMF *see* 2,5-dimethylfuran (2,5-DMF)
dried ginger 99–100
dry torrefaction 18

E-factor 139
economic feasibility 30, 31, 185
edible crops 11–13
electronic transmission microscopy 53
[emim][OAc] 172, 177
energy density, biofuels 149
entrainers *see* co-solvents
enzymatic hydrolysis 23, 55–6, 57
enzymatic reactions 18
    high-pressure reactors 72–3, 74
    pH and carbonic acid formation 71–2
    in supercritical CO_2 67–73
    temperature and pressure effects 70–1
    transesterification 28–9
    water content and 72

equations of state (EoS) 45, 49–50
esterification 67–9
transesterification 18, 23, 25, 28–9
ethane 48–9
ethanol
bioethanol 13, 14
as biofuel 149
cellulosic 53
CO_2 phase equilibrium 49–51
as co-solvent 4, 57–8
dehydration 119
in delignification 103–5
ethylene 2
Eucalyptus globulus 102–3
explosion treatment 21, 23
steam explosion 38–40, 53–4
extraction *see* supercritical CO_2 extraction

Fast Fourier Transformation pattern 142
fast pyrolysis 17
fatty acid methyl esters (FAMEs) 28, 76
fatty acids
fractionation 5
free fatty acid 68
hydrogenation 25
fermentation 13, 18
first generation biomass 11–13, 16, 84
Fisher-Tropsch process 17, 18, 22
flowrate 133
food *vs.* fuel competition 10, 84
formic acid 39
fossil fuels 9, 83, 145
Fourier transform infrared spectroscopy (FTIR) 53, 69, 70
fractionation of liquids 4–5
free fatty acid 68
fructose 121–2, 150
fuel *vs.* food competition 10, 84
fuels
fossil 9, 83, 145
jet fuels 10
see also biofuels
furfural viii, 25, 28
in the biorefinery concept 116–17
to 2-MF 155–6
in subcritical and carbonated subcritical water 127–30
in supercritical water 131
from xylans 120–1, 139
see also 5-hydroxymethyl furfural (5-HMF)
furfuryl alcohol 156

γ-valerolactone 21
gasification 17, 22
G1cN units 168
G1cNAc units 168
ginger bagasse 98–9
global energy consumption 9
global energy supply 10
global lignocellulosic production 85, 182
glucans 52, 86
glucose 121–3
aldol reactions 122
from cellulose 93
hydrolysis 123
reaction conditions 131
yield 27
see also sugar
1,2-glycerolcarbonate 183
green reaction medium 139–40
greenhouse gases 145
guaiacols 106–8
guayule 74

half-hydrogenated product 147–8
hardwood
chemical composition 85, 86
hydrolysis 94–5
hemicellulose 12, 13, 16, 167–8
anti-solvent conditions 171–4
chemical composition 85, 86, 138, 167
CO_2-assisted processes using 24
complexity and diversity of 183
hydrolysis 93–8, 117–23

Henry's law 19
hexane 146, 155
high-pressure batch reactors *see* batch reactors
high-pressure CO_2 and CO_2–H_2O systems 183, 185
 anti-solvent effect 169–78
 challenges of implementing 30
 chemical processes using 21–30
 essential features 19–20
 hydrolysis of biomass 88–90
 hydrolysis of biomass-derived polymers 91–110
 physical processes using 20–1
 principal benefits 31
 reaction configuration 132–3
 reaction medium and operating conditions 123–32
high-pressure continuous reactors *see* continuous reactors
high-pressure technology vii–viii, 181–6
Hildebrand solubility parameter 45
5-HMF *see* 5-hydroxymethyl furfural (5-HMF)
hops 2
hydrogels 169
hydrogen pressure 152, 155–6, 159
hydrogenation 18, 25–6, 28, 146–8
 catalytic transfer 150
hydrolysis 16, 18, 183
 acid 23, 24, 38–40
 autohydrolysis 38
 biomass 88–90
 biomass-derived polymers 91–110
 cellulose 26–7, 91–3, 117–23
 enzymatic 23, 55–6, 57
 hemicellulose 93–8, 117–23
 lignin 102–9
 mathematical models 118–19
 proteins 101–2
 reaction configuration 132–3
 reaction medium and operating conditions 123–32
 starch 98–100
 see also hydrothermal hydrolysis
hydrothermal gasification 17
hydrothermal hydrolysis 52, 93–4
 in depolymerization 106–9
 proteins 101–2
5-hydroxymethyl furfural (5-HMF) viii, 25, 28
 to alkane 145–9
 in the biorefinery concept 116–17
 to 2,5-DMF 149–55
 formation and decomposition of 121
 platform molecule 138–9
 in subcritical and carbonated subcritical water 128–9
 in supercritical water 131

imidazole 149
in situ synthesis
 catalyst characterization 142, 143, 144
 catalytic activity 142, 144–5
 general method 141–2
 phase behaviour studies 145
integrated biorefinery 10, 16
ionic liquids 58, 68–9
 2,5-DMF from 150
 and the anti-solvent process 172–8
 in delignification 106
 dissolving biopolymers 169–70
ionic product 5, 131
iron catalysts 17
isomerization 119–23

jet fuels 10

Kamlet-Taft parameters 175–6

lactic acid 116–17
levulinic acid viii, 25, 117

life cycle analysis 184
lignin 12, 13, 16
chemical composition 85, 86–7, 138
CO_2-assisted processes using 24
depolymerization 29–30
hydrolysis 102–9
recovery 54
valorization 183
Lignoboost® 54
lignocellulosic biomass 11–13, 16, 116
benefits of 84
chemical compositions 85–8
conversion problems 84
definition 137
global production 85, 182
see also biomass; cellulose; hemicellulose; lignin
lipases 28, 72
lipids
from algae 20, 77–8
CO_2-assisted processes using 25
phenolic 68
Lipozyme 68, 72
liquids
fractionation 4–5
see also ionic liquids; water

mass transfer limitations 28, 44, 140
mathematical models, hydrolysis 118–19
MCM-41
experimental methods 141–5
results 145–60
membrane reactors 73, 74
metal nanoparticles
experimental methods 141–5
results 145–60
methanation 18
methanol 48–9
5-methyl-2-furanyl methanol 154–5
2-methylfuran (2-MF) 28, 29, 139
from furfural 155–6
2-methyltetrahydrofuran 156
microalgae 11–14
functional properties and uses 77
lipid extraction 20, 77–8
microwave technology 100
mild pyrolysis 18
milling 133
Miscanthus X giganteus 103, 105
modifiers *see* co-solvents

naphtha 15
neutralization stage 54
nickel 17
nitrogen, removal of 149
non-edible crops 11, 12, 13–14
Novozym 435 69, 72, 75–6

oil production 12
operational conditions 123–32
operational costs 185
organic solvents 57, 102
in 5-HMF conversion 155
solubility parameters 45
organic wastes 13

palladium (Pd) catalyst 147, 150–2, 157
particle size reduction 17
Peng-Robinson EoS 49–50
1,5-pentanediol (1,5-PD) 139, 156–60
pentose 24
petroleum 9
oil production 12
refineries 14, 15
pH
CO_2, function of temperature and pressure 96
CO_2–H_2O system 19
enzymatic reactions and 71–2
and reaction severity 90
in subcritical water 124–5
phase behaviour
CO_2 41–7, 151–2
studies 145
THFA in CO_2 159

phase diagram 139–40
phase equilibrium 49–51
phenols 68, 106–9
phenylpropanoid units 86
phosphoric acid 39
physical processes
in biorefineries 17
in high-pressure CO_2 and CO_2–H_2O systems 20–1
Pinus taeda 103–4
π* solubility scale 175–6
platform molecules 13, 15, 138–9, 155
platinum catalyst 147
polyalcohols 24
polymerization
degree of 166, 168, 174
depolymerization 24, 29–30, 101, 106–9
ethylene 2
pre-treatment 16
biomass viii, 37–61
different methods 38–40
requirements for 38
precipitation mechanism 177–8
pressure
and the anti-solvent process 173–4
CO_2
and conversion 147–8
and conversion and selectivity 158–9
and product distribution 151–2
critical 5, 41–2
enzymatic reactions and 70–1
hydrogen 155–6
and pH of CO_2 96
pressure–composition isotherms 50–1
pressure–density diagrams 43
pressure–solubility parameter diagram 45–6
pressure–sound speed diagram 44
pressure–temperature and pressure–density projections 42
pressure–viscosity diagram 44
and solubility of CO_2 89–90, 124–5
see also high-pressure entries
pressurizing step 59
Promod 144 71
propane 4
proteins 16, 88
CO_2-assisted processes using 24
hydrolysis 101–2
purification 24
biodiesel 21, 25
pyrolysis 17–18, 22

reaction configuration 132–3
reaction feeding mode 59–60
reaction medium 123–32
green 139–40
subcritical and carbonated subcritical H_2O 124–30
supercritical H_2O 130–2
reaction rate 118–19
reaction severity 19, 90
reaction time *see* residence time
recycling, CO_2 133
red spruce wood 103–4
Renewable Energy Sources and Climate Change Mitigation 83
Renmatix 60
research octane number 149, 155
residence time 27, 101, 126, 131, 132
and the anti-solvent process 173–4
rhodium (Rd) catalyst 147, 157, 159–60
rice straw 85
ROSE (Residuum Oil Supercritical Extraction) 2
ruthenium 17
rye straw 128

scale-up, CO_2 pre-treatment 58–60
scanning electron microscopy (SEM) 53, 70, 105

second generation biomass 11–14, 16, 84
selectivity 145
  to alkane 147–8
  conversion and pressure in CO_2 158–9
  2,5-DMF 151, 153–5
  2-MF 156
  1,5-PD 157–8
semi-continuous reactors 132–3
separation procedures 17
severity factor 20, 53, 90, 125
silicon-MCM-41 (Si-MCM-41) 141–5
softwood, chemical composition 85, 86
solubility
  CO_2, function of temperature and pressure 89–90, 124–5
  π^* solubility scale 175–6
solubility parameter 45–7
solvatochromic effects 175–7
solvents *see* anti-solvent process; co-solvents; organic solvents; reaction medium
sorbitol 145–6
starch 16, 85
  chemical composition 87–8
  CO_2-assisted processes using 24, 26
  hydrolysis 98–100
  production 12–13, 16
steam explosion 38–40, 53–4
steam reforming 18
subcritical fluids, in pre-treatments 40
subcritical H_2O 124–30
  residence time 132
  use of CO_2 under 51–4
sugar 16
  from hemicelluloses 98–100
  production 12–14
  yield 74–5
  see also glucose
sugarcane bagasse 51–6, 58
  chemical composition 85
  delignification 103–4, 106
  hydrolysis 95, 97
sulfur, removal of 149
sulfur dioxide 53–4
supercritical anti-solvent effect (SAS) 166, 170–1
supercritical CO_2
  biomass conversion in 73–8
  biorefinery process based on 60
  enzymatic reactions in 67–73
  fractionation of liquids 4–5
  microalgal biomass in 77–8
  properties 66
  as reaction medium 139–41
  use of 54–6
    under subcritical water 51–4
  see also high-pressure CO_2 and CO_2–H_2O systems
supercritical CO_2 extraction
  early history 1–3
  physical processes using 20–1
  role of water in 3–4
supercritical fluids 2
  in pre-treatments 40–1
supercritical H_2O 5–6, 7
  oxidation 5–6
  reactions in 130–2
  residence time 132
  see also high-pressure CO_2 and CO_2–H_2O systems
supersaturation 170–1
sweet potato 98–9
switchgrass 128–9
  chemical composition 85
  hydrolysis 94–6
SWOT analysis 184
syngas 16, 17

tautomerization 119–23
technical readiness level (TRL) 184
temperature
  and the anti-solvent process 173–4
  conversion and product selectivity 153
  critical 1, 5, 41–3

temperature (*continued*)
enzymatic reactions and 70–1
and pH of CO_2 96
pressure–temperature projections in CO_2 42
and solubility of CO_2 89–90, 124–5
tetraethylorthosilicate 141–2
tetrahydro-5-methyl-2-furanmethanol (MHTFM) 151, 153–4
tetrahydrofuran (THF) 155
tetrahydrofurfuryl alcohol (THFA) 139, 156–60
thermochemical processes 17–18, 21, 26
thiophene 149
third generation biomass 11–14, 16
transesterification 18, 23, 25, 28–9
transformation rate *see* biomass conversion; conversion
transmission electron microscopy (TEM) 142, 144
triglycerides 12, 25
see also lipids
turmeric 99–100
turnover frequency 153

ultra-high-pressure processing 184
ultrasound 55, 56
unit operations 59

van der Waals, J.D. 2
van der Waals quadratic mixing rule 49–50
viscosity 43–5
volume of expansion 175

water
in 5-HMF conversion 154–5
content in enzymatic reactions 72
dielectric constant 48
as reaction medium 140–1
reactions of cellulose and hemicellulose in 119–23
solubility parameter 46
in supercritical CO_2 extraction 3–4
see also high-pressure CO_2 and CO_2–H_2O systems; subcritical H_2O; supercritical H_2O
wet torrefaction 18, 22
wheat straw 128
chemical composition 85
hydrolysis 95, 97

X-ray diffraction (XRD) 53–4, 142, 143
xylans 52, 86, 93–5, 120–1, 139
xylose 122–3, 129, 131